Professor Yoram Baram continues to revisit our understanding of the brain with bold, innovative ideas. In this groundbreaking book, he introduces, for the first time, a method by which one optimal neural circuit can guide action and then be seamlessly replaced by a more suitable one. This dynamic orchestration of neural activity enables successive movement in patients with neurological impairments, using principles as precise and intricate as those of cutting-edge computer science. The patient outcomes presented here represent just a glimpse of the remarkable improvements achieved under this approach.

—**Alberto J. Espay**, MD, MSc, FAAN, Professor of Neurology, Director and Endowed Chair, Gardner Family Center for Parkinson's Disease and Movement Disorders, University of Cincinnati Academic Health Center

The book Brain Proximity Measures, *suggested by Professor Yoram Baram as the underlying structure of brain activity, presents a powerful approach to the optimal choice of a neuronal circuit for motion control. Specifically, it selects a single neural circuit that maximizes the effectiveness and accuracy of a given group of neural circuits. As a result, the selected neural circuit produces the most suitable performance required. The mathematical details are presented in the opening chapters of this book for the mathematically inclined. Subsequent chapters present actual results obtained by Prof. Baram and his medical collaborators in testing neurologically impaired patients. The results show highly improved performance. The unique contribution of this book, rooted in Prof. Baram's original PhD thesis, is the idea that the best motion performance is created by neuronal selection. Specifically, the transition between groups of multiple neural circuits, to a single, most suitable neural circuit, is fast, resulting in the highest medical improvement.*

—**Uri Shaked**, Professor, Tel Aviv University, Fellow of the IEEE, Fellow of the Institute of Mathematics and its Applications (IMA), Winner of the 2017 Israel Prize for Engineering Research

In Brain Proximity Measures, *Yoram Baram presents a groundbreaking synthesis of control theory, neuroscience, and rehabilitation medicine to explore how the brain selects and activates neural circuits for motion control, particularly in neurologically impaired individuals. Drawing on decades of work in mathematical modelling and inspired by personal encounters with Parkinson's patients, Baram introduces the concept of neural proximity selection: the brain, when lacking a perfect circuit for a given task, selects the one closest in functional space. This insight leads to the central idea that motion controlled by impaired brains can be enhanced by guiding the system to better approximate optimal circuit activation — an idea tested through mathematical rigor and confirmed in clinical trials using wearable sensory-feedback devices. Merging theory and practice, the book offers both a formal framework for understanding stochastic convergence and model selection in neural systems, and an accessible presentation of experimental results across Parkinson's disease, stroke, cerebral palsy, and multiple sclerosis. It challenges conventional rehabilitation approaches by advocating closed-loop, autonomous sensory entrainment over externally paced therapies. It is a must-read for neuroscientists, clinicians, and engineers alike — an inspiring demonstration of how theoretical insight can directly improve human lives.*

—**Dario Ringach**, Professor of Neurobiology & Psychology, David Geffen School of Medicine, University of California, Los Angeles

BRAIN PROXIMITY MEASURES

Selecting the Optimal Neural Circuit from a Set

Yoram Baram

Technion — Israel Institute of Technology, Israel

World Scientific

EW JERSEY • LONDON • SINGAPORE • BEIJING • SHANGHAI • HONG KONG • TAIPEI • CHENNAI • TOKYO

Published by

World Scientific Publishing Co. Pte. Ltd.
5 Toh Tuck Link, Singapore 596224
USA office: 27 Warren Street, Suite 401-402, Hackensack, NJ 07601
UK office: 57 Shelton Street, Covent Garden, London WC2H 9HE

Library of Congress Control Number: 2025041970

British Library Cataloguing-in-Publication Data
A catalogue record for this book is available from the British Library.

BRAIN PROXIMITY MEASURES
Selecting the Optimal Neural Circuit from a Set

ISBN 978-981-98-1926-3 (hardcover)
ISBN 978-981-98-1927-0 (ebook for institutions)
ISBN 978-981-98-1928-7 (ebook for individuals)

For any available supplementary material, please visit
https://www.worldscientific.com/worldscibooks/10.1142/14481#t=suppl

Desk Editors: Nambirajan Karuppiah/Lai Fun Kwong

Typeset by Stallion Press
Email: enquiries@stallionpress.com

I am deeply indebted to Ophira, Tami, Orli, Amir, Ur, Maya, Lotem, Sara, Ya'ar and Arava for being there.

Prologue

Following my PhD in electrical engineering and computer science at MIT in 1976, I devoted the next 20 years to advanced control theory, addressing and teaching relevant mathematical concepts, with a side interest in aircraft control. My coincidental acquaintance with a Parkinson's patient has turned into a passionate desire to help people with walking impairment. My first article on this topic, "Walking on Tiles", was published in 1999. In time, I tested patients with Parkinson's disease, multiple sclerosis, idiopathic senile gait disorders, brain stroke and cerebral palsy. For this purpose, I developed a portable visual and auditory wearable device, which was approved by the US Patent Office in 2004. Following a sequence of experiments, conducted first with some of my students and then with world-renowned medical professors, my device was found to be of significant help for patients with movement disorders. The results of these experiments were published in leading medical journals. At the same time, I kept my interest in conducting scientific studies of a variety of cortical aspects. These studies were mostly addressed by relevant mathematics and tested by computational means.

It was my recent realization that my PhD thesis, concluded many years ago, was largely based on two main chapters, one termed "Stationary Linear Systems", while the other termed "Non-Stationary Linear Systems". Both chapters, originally motivated by the concept of advanced aircraft control, were now based on finite collections of neural circuits. In natural terms, the strongest neural circuit in a group brings down the other neural circuit to elimination. Yet, subject to another mission, another neural circuit of the group may recover, taking over a new responsibility. A reasonable assumption is that, in the absence of a perfectly accurate neural circuit for a given mission, the brain selects the most suitable neural circuit, eliminating the other neural circuits in the group. For the neurologically impaired, such transformations of a neural brain group, which I termed "Brain Proximity Measures", can save a patient from disastrous walking results. A particular aspect of neural circuits shown in this book is mathematical irreversibility. Specifically, the exact activity of a neural circuit cannot be repeated (see Chapter 7 for a mathematical explanation).

While the first seven chapters of this book provide the mathematical details supporting the medical background, some of these chapters require a high mathematical familiarity. On the other hand, the relatively large variety of medical test results reported in the last 10 chapters of this book are relatively easy to derive and explain. The brain proximity measures are meant to represent a finite set of neural networks, one of which is selected at a time for specific performance. Such performance may be associated with a given patient. In other words, the patients' active neural circuit is helped by external support, which is provided by the attached patented audio-visual device. In patient's motion reality, the corresponding medical device plays the role of the relevant brain proximity measure. More deliberate constitutions are described in the corresponding chapters.

Yoram Baram, Professor Emeritus
Roy Matas/Winnipeg Chair in Biomedical Engineering
Computer Science Department
Technion, Israel Institute of Technology

Contents

Chapter 1

Introduction

1.1 Some Preliminaries

Originating in a desire to maximize accuracy and safety in assigning mathematical models to physical observations, a fundamental question arises when the selection of a model from a given set of models presents a dilemma. The situation is complicated by the fact that the relationship between the observation and the sought mathematical model is uncertain and can only be specified in a probabilistic framework. For mathematical tractability, the problem is formulated as one of selecting, via some criterion, the "best" model from a specified model set. The formulation of the mathematical problem requires, then, the choice of a model set, on the one hand, and the choice of a model selection criterion, on the other. The first choice presents an obvious tradeoff. The more strictly the model set is specified, the more tractable is the mathematical solution, but the less probable is the case that a correct model is included in the specified model set. As an illustration, consider the two extreme situations. If the model set consists of a single model, then the selection is trivial, but the model may not be an adequate representative of the observed phenomenon. On the other hand, if the model set is the abstract "set" of "all models", then it obviously contains the correct model, but a mathematical solution (or formulation) of the model selection problem is then not feasible.

The model set can be naturally specified in terms of a parameter set, such that to each parameter there corresponds a model and vice versa. The terms *model set* and *parameter set* will be used interchangeably, and precise relationships between them will be definable. The model selection problem can then be naturally defined as a parameter estimation problem. Given a parameter set, the problem formulation requires the selection of a parameter estimation criterion. The true parameter cannot, in general, be assumed to belong to the prespecified parameter set, as asserted above. It turns out the maximum likelihood estimate, defined in Chapter 2, is most adequate for this situation. On the other hand, the Bayesian methods of maximum *a posteriori* probability and least squares, also defined in Chapter 2, intrinsically assume that the true parameter is a member of the parameter set.

One objective of our study is to provide in a very general setting answers to the following questions: under what conditions do the maximum likelihood and the Bayesian estimates converge to some parameter in the parameter set? What distinguishes the selected model from the other models in the model set, and what is its relationship to the true model? For the selection of the estimation procedure, it is reasonable to assume that the true parameter is a member of the set when it is not? Is the true model selected when it is a member of the model set? A question that arises naturally in this setting is: what is the best approximation of a complex model by a simple one?

A particular problem of considerable practical significance is that of dynamic system identification. The situation described above, and the questions raised, naturally apply to the system identification problem. We analyze the asymptotic behavior of system identification procedures in the presence and absence of the true model in a given model set. The analysis also suggests a systematic approach to certain system modeling problems of practical significance.

A certain part of the analysis in this book is restricted to the case where the model set is finite. This restriction serves several purposes. We chose to emphasize the statistical properties of the observation sequences involved (such as their content of information) and to avoid considerations of topological conditions on the parameter set, which are unavoidable if results for, e.g., infinite compact parameter sets are desired. This makes the analysis considerably simpler and enables us to consider very general classes of observation sequences. It is nevertheless demonstrated that the results obtained for finite parameter sets may be extended to compact sets by additional requirements on the topology of the set, such as uniform continuity of the density functions involved.

In addition to the above considerations, the case of finite parameter sets has a considerable practical significance as a method of approximation. Identification techniques for finite sets of models are considerably faster than those for infinite sets, as the search procedure for the parameter satisfying the estimation criterion is practically trivial. In fact, our analysis makes a strong case for the finite model set, taking the viewpoint that the true model is, in most cases, not included in any prespecified set of models. Identification is thus a procedure of finding an approximate model, whether a finite or an infinite model set is considered. The approximation is nevertheless "coarser" when fewer models are included in the model set.

It should, however, be emphasized that a substantial portion of the present study applies to parameter sets that may be infinite or even non-compact. This is the case in the derivation of distance measures on the parameter set and the consideration of system modeling problems.

For comparison with earlier results, we note that the convergence of the parameter estimates is considered in the probabilistic senses of convergence almost everywhere (a.e.) and convergence in the mean square (m.s.), which will be defined in the following chapter. Consistency is traditionally defined as the convergence a.e. of the estimate to the true parameter when it is included in the parameter set.

1.2 Historical Review

Parameter estimation techniques have been studied ever since the introduction of the maximum *a posteriori* probability (MAP), the least-squares criteria by Gauss (1809) and Laplace (1820), and their later studies by Edgeworth (1908). Fisher (1922) proposed the maximum likelihood (ML) estimate, which has since gained considerable popularity due to its intuitive appeal and its asymptotic properties (e.g., LeCam (1953)).

The consistency of ML estimates for sequences of independent and identically distributed (i.i.d.) observations was proved by Cramer (1946), who assumed differentiability to fourth order of the probability density function involved. Differentiability assumptions were dispensed with in proofs by Doob (1934) and Wald (1949). The main tool in proving consistency for i.i.d. observations is, naturally, the strong law of large numbers. Roussas (1965) proved the consistency of ML estimates for the case of ergodic Markov observation sequences, employing the ergodic theorem. The mean square (MS) convergence of LS estimates given i.i.d. observations was considered by Liporace (1971), who showed, via the multiplication rule for independent random variables, that the MS error of these estimates is exponentially diminishing. In the case where the true parameter is not included in the parameter set, the estimates were shown to converge to a parameter in the set, which is most similar to the true parameter. The measure of similarity suggested by Liporace is related to the information measures introduced in this book. Caines (1975a) proved and applied the submartingale property of sequences of maximum ratios on finite parameter sets to prove the consistency of ML estimates on such sets for the general class of observation sequences, satisfying a certain probabilistic condition. Baram and Sandell (1978) extended Caines' results to Bayesian estimates, which were shown to be consistent a.e. and in the mean square, and showed that Caines' condition applied to stationary Gaussian linear systems.

The identification of linear dynamical systems employing parameter estimation techniques has been studied intensively for decades. However, several consistency proofs that have appeared in the early literature have overlooked the fact that, for consistent estimation on compact parameter sets, uniform convergence of the associated probability densities on the parameter set is necessary, while pointwise convergence only provides consistency for finite parameter sets. Correct consistency proofs have appeared in the literature in recent years. Caines and Rissanen (1974) (see also Rissanen and Caines (1974)) proved the consistency of the ML estimates for autoregressive and moving average (ARMA) observation sequences. Lejung proved the consistency of stochastic approximation techniques (1974a) and the consistency of a class of prediction error techniques (1974b) (see also Lejung (1975)). Caines (1975b) proved the consistency of stationary processes of a more general class of prediction error techniques, which includes the maximum likelihood technique for the case of stationary Gaussian observation sequences. The topological requirements specified by Caines (1975b) reduce in the finite parameter set case to a requirement that there exists a one-to-one correspondence between the parameter set and the set of the system's impulse responses corresponding to the system's innovations representation. Similar conditions were suggested by Tse and Weinert (1975) (see also Tse (1976)) and by Hawkes and Moore (1976) (see also Moore and Hawkes (1974)), who considered the convergence of Bayesian estimates on finite sets of stationary Gaussian linear systems. The condition suggested by Baram and Sandell (1976) is a uniqueness condition on the output statistics associated with the different models in the model set. Other statistical conditions are motivated and derived in this book.

Information methods have been suggested by many authors for the solution of the related problems of hypothesis testing, signal selection and model identification. Kullback's information distance measure (1959) has proved to be useful in the analysis of parameter estimation and model identification techniques. Kullback's information measure was employed by Liporace (1971), and, following Liporace, by Hawkes and Moore (1976) in their studies of parameter estimates, given i.i.d. and stationary Gaussian observations. In this book, we define and employ information measures, which prove to possess valuable properties lacking in Kullback's information measure, such as the metric property on the parameter space. Other information measures, defined and employed in the literature, will be mentioned in the following chapter as they are compared with the information measures defined in the remainder of this book.

Part I

Convergence on Multiple Parameter Sets

Chapter 2

Probability Spaces and Parameter Estimates in Stochastic Convergence

The purpose of this chapter is to present the underlying mathematical setup for this book and recall definitions and results from probability and estimation theory that will be used in the following chapters.

Since a major objective of this book is to analyze, using correct assumptions, parameter estimates that may be based on incorrect assumptions, it is essential to define at the outset the different probabilistic frameworks in which the estimates are defined and the analysis is performed. We first introduce the correct framework in which the analysis is performed. It consists of an underlying probability space and a separate parameter space, of which the true parameter may or may not be a member. Likelihood ratios and maximum likelihood estimates are naturally defined in this framework. On the other hand, Bayesian parameter estimates are defined in a different framework where the parameter space is a part of the underlying sample space. Consequently, the existence of a probability measure defined on the parameter space (i.e., assigning to each set in the parameter space the probability that it includes the true parameter) is postulated. The Bayesian framework then inherently includes the assumption that the true parameter is a member of the given parameter space and is inadequate for the analysis of the general case considered in this study. Thus, while the Bayesian setup is assumed in the definition of Bayesian estimates, the analysis of these estimates as well as the maximum likelihood estimates is performed using the underlying, non-Bayesian framework.

Readers unfamiliar with the notion of measure and probability spaces may identify here, and in the following chapters, the functions $f(Z^n)$, $f(z_n|Z^{n-1})$ and $f(s|z^n)$, with the familiar probability density, conditional probability density and *a posteriori* probability density functions on Euclidean observations and parameter spaces. Several symbols and terms, mostly standard in probability and information theory, are introduced in this chapter.

2.1 Observations, Parameters and Likelihood Ratios

Consider a measurable space (Ω, U), where Ω is some sample space and U is a σ-algebra of subsets of Ω. The observation sequence (z_n) is a stochastic process on a probability space (Ω, U, P_*), with values in a measurable space (D, ∂), called the observation space. We shall be interested in the case $(D, \partial) = (R^\ell, B^\ell)$, where R^ℓ is the ℓ-dimensional Euclidean space and B^ℓ is the σ-algebra of Borel sets in R^ℓ (a Borel set being any subset of a topological space that can be formed from its open sets through the operations of countable unions). We call P_* the true measure and $*$ the true parameter.

The parameter space S is a set such that for each $s \in S$, there exists a probability measure P_s defined on (Ω, U). Let $T \equiv (* \cup S)$. Obviously, $* \in T$, but $*$ need not belong to the set S.

For each $s \in T$, we denote by E_s expectation taken with respect to P_s. We use the notation a.e. (almost everywhere) to denote events of P_* measure one. Events of P_s measure one will be denoted a.e. P_s.

Recall that the conditional expectation of a random variable x on (Ω, U, P) given $A \in U$ is a U-measurable random variable denoted $E^A(x)$ such that

$$EE^A(x) = E(x) \tag{2.1}$$

For each $s \in S$, we shall denote by E_s^A the conditional expectation given A, taken with respect to P_s.

If μ and ν are measures defined on (Ω, U), then μ is said to be absolutely continuous with respect to ν if for any set $A \in U$, $\nu(A) = 0$ implies $\mu(A) = 0$. μ is said to be singular with respect to ν if it is not absolutely continuous with respect to ν.

Let $(U_n) \equiv (U_n(z^n))$ be the increasing family of σ-subalgebras of U generated by

$$z^n \equiv (z_1, \ldots, z_n) \tag{2.2}$$

For each $s \in T$ and for each $n \geq 0$, let $P_{s,n}$ denote the restriction of P_s to U_n. Suppose that for each $n \geq 0$, the measure $P_{s,n}$ are absolutely continuous with respect to some measure λ_n defined on (Ω, U). Then

$$f_{s,n} \equiv \frac{dP_{s,n}}{d\lambda_n}; \quad s \in T \tag{2.3}$$

are the Radon–Nikodyn derivatives (or densities) between the respective measures. The likelihood ratio between two parameters s, $t \in T$ is defined as

$$h^s_{t,n} \equiv \frac{dP_{s,n}}{dP_{t,n}} = \frac{f_{s,n}}{f_{t,n}} \tag{2.4}$$

provided that $P_{s,n}$ is absolutely continuous with respect to $P_{t,n}$. When the time parameter n is included in the argument, we shall use the somewhat shorter notation:

$$f_s(a(n)) \equiv f_{s,n}(a(n)); \quad s \in T; \quad a(n) \in U_n$$

in particular

$$f_s(Z^n) \equiv f_{s,n}(Z^n); \quad s \in T \tag{2.5}$$

$$h^t_s(Z^n) \equiv h^t_{s,n}(Z^n); \quad s, t \in T \tag{2.6}$$

For any $c \in U_n$ and $b \in U_n$ such that $f_{s,n}(b) \neq 0$ for all $s \in S$, the conditional densities of c given b are

$$f_{s,n}(c|b) \equiv \frac{f_{s,n}(c, b)}{f_{s,n}(b)}$$

in particular

$$f_s(z_n|Z^{n-1}) = \frac{f_s(Z^n)}{f_s(Z^{n-1})}; \quad s \in T \tag{2.7}$$

The conditional likelihood ratios are then defined as

$$h_t^s(z_n|Z^{n-1}) \equiv \frac{f_s(z_n|Z^{n-1})}{f_t(z_n|Z^{n-1})}; \quad s\,t \in T \tag{2.8}$$

for any z^n such that $f_t(z_n|Z^{n-1}) \neq 0$ for all $t \in T$.

The following condition will be assumed throughout the book:

(c2.1) For all $s \in S$, the probability measures $P_{s,n}$ are mutually absolutely continuous.

2.2 Bayesian Probability Densities

Consider a measurable space (Ω, U),where Ω is some sample space and U is a σ-algebra of subsets of Ω, and a measurable space (S, U^s), where S is the parameter space and U^s is a σ-algebra of subsets of S. Let (Ω^b, U^b) be a measurable space, where

$$\Omega^b \equiv \Omega \times S$$

and

$$U^b \equiv U \times U^s$$

are the Cartesian products of the respective sample spaces and σ-algebras. Let P^b be a measure on (Ω^b, U^b). We denote by E^b expectation and by E^{b^A}; $A \in U$ conditional expectation given A, taken with respect to P^b. We call the restriction P_0^b of P^b to (S, U^s) the *a priori* probability measure on (S, U^s). Suppose that P_0^b is absolutely continuous with respect to some measure ν_0 *on* (S, U^s), then the density

$$f_0^b = \frac{dP_0^b}{d\nu_0} \tag{2.9}$$

is well defined. In particular, we call

$$f_0^b(s); \quad s \in S \tag{2.10}$$

the *a priori* probability density on S with respect to the measure ν_0.

Let (z_n) be a stochastic process on the probability space (Ω^b, U^b, P^b) with values in a measurable space (D, ∂), and let $(U_n^b) \equiv (U_n^b(Z^n))$ be the increasing sequence of σ-subalgebras of U^b, generated by $z^n \equiv (z_1, \ldots, z_n)$. Let P_n^b; $n \geq 1$ be the restriction of P^b to U_n^b, and for each $n \geq 1$, let P_n^b be absolutely continuous with respect to some measure ν_n defined on (Ω^b, U^b). Then the density

$$f_n^b \equiv \frac{dP_n^b}{d\nu_n} \tag{2.11}$$

is well defined. We shall be particularly interested in the *a posteriori* probability density of s, given Z^n

$$f^b(s|Z^n) \equiv f_n^b(s|Z^n) \equiv \frac{f_n^b(s, Z^n)}{f_n^b(Z^n)}; \quad n \geq 1 \tag{2.12}$$

assuming $f_n^b(Z^n) \neq 0$.

Let the parameter set S be finite, i.e.,

$$S = \{s_j;\ j \in K \equiv (0, \ldots, p)\} \tag{2.13}$$

For each $j \in K$, let

$$1_{s_j}(s) \equiv \begin{cases} 1 & s = s_j \\ 0 & s \neq s_j \end{cases}$$

Then

$$\nu_0 \equiv \sum_{j=0}^{p} 1_{s_j}(s)$$

is a measure ("the counting measure") on (S, U^s). Let λ be a measure on (Ω, U), then $\nu_n = \nu_0\lambda$; $n \geq 1$ is the product measure on (Ω^b, U^b). Suppose that P_n^b is absolutely continuous with respect to ν_n (i.e., the entire measure P_n^b is concentrated on the set $\Omega \times \{s_i; i \in K\}$)) for all $n \geq 0$, then we have

$$f_0^b(s) = \sum_{i=0}^{p} f_0^b(s_i) 1_{s_i}(s) \tag{2.14}$$

and

$$f_n^b(s, z^n) = \sum_{i=0}^{p} f_n^b(s_i, z^n) 1_{s_i}(s) \tag{2.15}$$

Hence,

$$f_n^b(Z^n) = \int_S f_n^b(s, Z^n) d\nu_0 = \sum_{i=0}^{p} f_n^b(s_i, Z^n) = \sum_{i=0}^{p} f_0^b(s_i) f^b(Z^n|s_i) \tag{2.16}$$

where we have applied Bayes' rule

$$f_n^b(s_i, z^n) = f_0^b(s_i) f_n^b(z^n|s_i) \tag{2.17}$$

Substituting (2.16) and (2.17) into (2.12) yields for each $j \in k$

$$f^b(s_j|z^n) = \frac{f_0^b(s_j) f_n^b(Z^n|s_j)}{\sum_{i=0}^{p} f_0^b(s_i) f_n^b(Z^n|s_i)} \tag{2.18}$$

Note that

$$f_n^b(Z^n|s_j) = f_{s_j}(Z^n) \tag{2.19}$$

where the right-hand side is defined by (2.5). Thus, finally, for the finite parameter set S,

$$f^b(s_j|Z^n) = \frac{f_0^b(s_j) f_{s_j}(Z^n)}{\sum_{i=0}^{p} f_0^b(s_i) f_{s_i}(Z^n)} \tag{2.20}$$

2.3 Parameter Estimates and Stochastic Convergence

An estimate $\hat{s}_n$ on S is a U_n- measurable mapping from Ω onto S.

A maximum likelihood (ML) estimate on S is an estimate $\hat{s}_n \in S$ such that

$$\{f_s(Z^n);\ s \in S\} \leq f_{\hat{s}_n}(Z^n)$$

A maximum *a posteriori* probability (MAP) estimate on S is an estimate $\hat{s}_n \in S$ such that

$$\{f^b(s|z^n); s \in S\} \leq f^b(\hat{s}_n|Z^n)$$

Let S be linear. Then, a least-squares (LS) estimate on S is an estimate $\hat{s}_n \in S$ such that

$$E^b\{(s'_n - *)^T(\hat{s}'_n - *)\} \geq E^b\{(\hat{s}'_n - *)^T(\hat{s}'_n - *)\}$$

for any estimate s'_n on S. x^T denotes x transposed and $*$ denotes the true parameter, assumed to have the same dimension as s.

Let the true parameter be assumed to belong to a finite set $\{s_j \in R^m; j \in k\}$. Then the LS estimate on R^m at instance n is the conditional expectation

$$E^{b^{U_n}} s = \int_s s f^b(s|Z^n) d\nu_0 = \sum_{j=0}^{p} s_j f^b(s_j|Z^n) \qquad (2.21)$$

A stochastic sequence (x_n) on (Ω, U, P) is said to converge almost everywhere (a.e.) to a random variable (x_n) on (Ω, U, P) if

$$\lim_{n\to\infty} x_n = x \quad \text{a.e.}$$

A stochastic sequence (x_n) on (Ω, U, P) is said to converge in the mean (or in L_1) to a random variable (x) on (Ω, U, P) if

$$\lim_{n\to\infty} E|x_n - x| = 0 \quad \text{a.e.}$$

A vector-valued stochastic sequence (x_n) on (Ω, U, P) is said to converge in the mean square (in M.S. or in L_2) to a random vector (x) on (Ω, U, P) if

$$\lim_{n\to\infty} E|x_n - x|^2 = 0$$

where $|x| \equiv (x^T x)^{1/2}$.

A sequence of parameter estimates $(\hat{s}_n)$ is said to be consistent a.e. or in the mean square if it converges a.e. or in the mean square to the true parameter.

We now present without proofs three well-known results from probability theory, which are used in this book.

Theorem 2.1 (Jensen's inequality (Jensen, 1906; Bauer, 1972, p. 322)). *Let x be a real integrable random variable on a probability space (Ω, U, P) with values in R^1, and let $g(x)$ be a convex integrable function on R^1, then*

$$g(Ex) \leq Eg(x)$$

Theorem 2.2 (Fatou's Lemma (Bauer, 1972, p. 71)). *Let (x_n) be an integrable stochastic sequence on (Ω, U, P) such that $x_n \geq 0$ a.e. for all n, then*

$$E \lim_{n\to\infty} \inf x_n \leq \lim_{n\to\infty} \inf E\, x_n$$

Theorem 2.3 (Lebesgue's dominated convergence theorem (Chung, 1974, p. 421)). *Let (x_n) be an integrable stochastic sequence on (Ω, U, P). Then if*

$$\lim_{n\to\infty} x_n = x \quad \text{a.e.}$$

where x is an integrable random variable on (Ω, U, P) and if there exists some integrable random variable y on (Ω, U, P) such that

$$Ey < \infty$$

and

$$|x_n| \leq y \text{ a.e.} \quad for\, all\, n$$

then

$$\lim_{n\to\infty} Ex_n = Ex$$

2.4 Martingales and Martingale Difference Sequences

Let (Ω, U, P) be a probability space and let (U_n) be an increasing family of $\boldsymbol{\sigma}$-subalgebras in $\boldsymbol{U}$. A U_n-measurable stochastic sequence(x_n) on (Ω, U, P) is called a U_n-martingale if for each n

(a) $E|x_n| < \infty$
(b) $E^{U_{n-1}} x_n = x_{n-1}$ a.e.

If the equality in (b) is replaced by $\leq$, then (x_n) is called a U_n-supermartingale.

It can be shown (Doob, 1953) that the likelihood ratio sequences $\left(\frac{dP_{s,n}}{dP_{*,n}}\right)$, $s \in S$, defined in Section 2.2 are U_n-martingales according to the measure P_*. Hence,

$$E_*^{U_{n-1}} \frac{dP_{s,n}/dP_{s,n-1}}{dP_{*,n}/dP_{*,n-1}} = \frac{dP_{*,n-1}}{dP_{s,n-1}} E_*^{U_{n-1}} \frac{dP_{s,n}}{dP_{*,n}} = \frac{dP_{*,n-1}}{dP_{s,n-1}} \frac{dP_{s,n-1}}{dP_{*,n-1}} = 1$$

Consequently, we have by Eqs. (2.3), (2.7) and (2.8)

$$E_*^{U_{n-1}} h_*^s(z_n|Z^{n-1}) = E_*^{U_{n-1}} \frac{f_s(z_n|Z^{n-1})}{h_*^s(z_n|Z^{n-1})} = 1 \quad \text{for each } s \in S \tag{2.22}$$

Theorem 2.4 (The martingale convergence theorem (Chung, 1974, p. 334; Bauer, 1972, pp. 341–343)). *Let (x_n) be a U_n-martingale on (Ω, U, P) and let $\sup_{n\geq 0} E\, x_n^+ < \infty$, where $x_n^+ = \sup(x_n, 0)$.*

Then (x_n) converges a.e. to a finite limit.

Let (Ω, U, P) be a probability space and let (U_n) be an increasing family of σ-subalgebras in U. A U_n-measurable stochastic sequence (x_n) on (Ω, U, P) is called a

U_n-martingale difference sequence if it is integrable and if

$$E^{U_{n-1}} x_n = 0 \text{ a.e.}$$

Let y_n be a stochastic sequence on (Ω, U, P) and let (U_n) be a sequence of σ-subalgebras of U, generated by $(y_1, \ldots, y_n)$. Then, clearly

$$(y_n - E^{U_{n-1}} y_n)$$

is a U_n-martingale difference sequence. Also, note that if (x_n) is a U_n-martingale difference sequence, then

$$x_n \equiv \sum_{m=1}^{n} x_m$$

is a martingale. Indeed,

$$E^{U_{n-1}} x_n = \sum_{m=1}^{n} E^{U_{n-1}} x_m = \sum_{m=1}^{n-1} x_m + E^{U_{n-1}} x_n = \sum_{m=1}^{n-1} x_m = X_{n-1}$$

2.5 Stationarity and Ergodicity

The purpose of this section is to provide definitions and convergence results for ergodic sequences, which will be used in this book. It is not intended to provide an elaborate presentation of the concept of ergodicity. For a thorough development of ergodic theory, the reader is referred to the works by, e.g., Doob (1953), Halmos (1956) and Chacon and Orenstein (1959).

Consider a probability space (Ω, U, P). A transformation from Ω to U is said to be the measure preserving if $P(T^{-1}A) = P(A)$ for all $A \in U$.

Given a measure-preserving transformation T, a U-measurable event A is said to be invariant if

$$T^{-1}A = A$$

Let (x_n) be a stochastic sequence on (Ω, U, P) with values in (R^ℓ, B^ℓ), where R^ℓ is the ℓ-dimensional Euclidean space and B^ℓ is the σ-algebra of Borel sets of R^ℓ. Let B^ℓ_∞ be the σ-algebra of Borel sets of R^ℓ_∞, where $R^\ell_\infty \equiv R^\ell \times R^\ell \times \cdots$ Then (x_n) is said to be stationary if for $k \geq 1$

$$P[(x_1, \ldots, x_n) \in C] = P[(x_{k+1}, x_{K+2}, \ldots,) \in C]$$

for every $C \in B^\ell_\infty$.

A stationary sequence (x_n) on (Ω, U, P) is said to be ergodic if every invariant event in U has probability zero or one. It can be shown (e.g., Stout, 1974, p. 168) that (x_n) is generated by a measure-preserving transformation (the shift operator), i.e.,

$$x_n(\omega) = x_{n-1}(T\omega) \tag{2.23}$$

Let (x_n) be a vector-valued stochastic sequence from (Ω, U, P) into (R^ℓ, B^ℓ) such that the probability density with respect to the Lebesgue measure on (R^ℓ, B^ℓ) of (x_n) is Gaussian on R^ℓ, with $Ex_n = m_x$, constant for all n

and

$E\{(x_n - m_x)(x_{n+k} - m_x)^T\}$ depends only on k.

Then (x_n) is a stationary Gaussian sequence.

Proposition 2.1 (Grenander, 1959, pp. 257–260; Doob, 1953, p. 494). *A zero-mean stationary Gaussian process is ergodic if and only if*

$$\lim_{n\to\infty} \frac{1}{n+1} \sum_{k=0}^{n} |R(k)|^2 = 0$$

where

$$R(k) \equiv E\{x_n x_{n+k}^T\}$$

and where $|R(k)|$ denotes the determinant of $R(k)$.

Theorem 2.5 (The ergodic theorem (Doob, 1953, p. 464; Halmos, 1956, p. 22; Weiner, 1949, p. 16)). *Let (x_n) be an ergodic sequence on (Ω, U, P) and let $f(x_n)$ be a U-measurable function such that $E|f(x_0)|$ is finite, then*

$$\lim_{n\to\infty} \frac{1}{n+1} \sum_{j=0}^{n} f(x_j) = Ef(x_0).$$

The following version of the central limit theorem of probability will prove useful in later chapters.

Theorem 2.6 (Billingsley, 1961). *Let (x_n) be an ergodic stochastic process on (Ω, U, P) such that $E(x_1^2)$ is finite and*

$$E^{U_{n-1}} x_n = 0 \quad \text{a.e.}$$

(i.e., (x_n) is an ergodic martingale difference sequence). Then the distribution of $n^{-1/2} \sum_{k=1}^{n} x_k$ approaches the Gaussian distribution with mean zero and variance $E(x_1^2)$.

2.6 Metric Spaces and Stochastic Metrics

Consider a set S and a real-valued function e on $S \times S$ which satisfies the following:

(i) $e(s; s) = 0$ for any $s \in S$
(ii) $e(s; t) = e(t; s)$ for any $s, t \in s$

(iii) $e(s;t) \leq e(s;r) + e(r;t)$ for any $s, t, r \in S$.
Then e is called a pseudo-metric on S. If, in addition to (i), (ii) and (iii), e satisfies

(iv) $e(s;t) = 0; s, t \in S$ implies $s \in t$
then e is called a metric on S. The pair (S, e) is called a metric space.

Now, consider a probability space (Ω, U, P) and an increasing family (U_n) of subalgebras of U. Let (e_n) be a (U_n)-measurable sequence of functions on $S \times S$ such that each e_n satisfies (i)–(iii) above. Then we shall call (e_n) a stochastic pseudo-metric sequence on S. If each e_n satisfies (i)–(iii) above, we shall call e_n a stochastic metric sequence on S.

Chapter 3

Information

In this chapter, we develop the notion of information in a sequence of observations, favoring one parameter in a given parameter set against another. We do not make the assumption, common in the derivation of other information measures in information theory, that the true parameter is included in a known set, or, equivalently, that the true measure belongs to a known set of measures. The mean and the conditional mean values of the discriminating information in a single observation are shown to possess properties that will prove useful in the following chapters. In particular, their absolute values are metrics, or distance measures, of the parameters and the parameter space. This provides a meaningful measure of the relative closeness of parameters to the true parameter. The new information measures are then compared with other measures common in information theory.

3.1 The Information in a Single Observation

Let S be a parameter space and let $T = ({}^*\cup S)$, where * is the true parameter. If, for some pair of parameters $s, t \in T$

$$f_s(z^n) > f_t(z^n)$$

or, equivalently,

$$\log f_s(z^n) > \log f_t(z^n)$$

we say that the parameter s is favored over the parameter t by the observations z^n. Then $\log f_s(z^n)$ may be regarded as a measure of the information in z^n for selecting a parameter from the set T. The difference

$$\log f_s(z^n) - \log f_t(z^n) = \log_t^s(z^n) \tag{3.1}$$

is then a measure of the information in z^n for selecting between s and t. If Eq. (3.1) is positive, then s is favored, and if it is negative, then t is favored. The difference

$$\log h_t^s(z^n) - \log f_t^s(z^{n-1}) = \log h_t^s(z_n|Z^{n-1}) \tag{3.2}$$

is then a measure of the difference between the information favoring s against t at instant n and the information favoring s against t at instant $n-1$. It can then be regarded as a measure of the information favoring s against t in the observation z_n. We define

$$I_n(s;t) \equiv E_*^{U_{n-1}} \log h_t^s(z_n|Z^{n-1}) \tag{3.3}$$

as the conditional mean information in z_n favoring s against t and

$$\overline{I}_n(s;t) \equiv E_* \log h_t^s(z_n|Z^{n-1}) \tag{3.4}$$

as the mean information in z_n favoring s against t. (A more general form of (3.3) would be $\tilde{I}_n(s;t) \equiv E_*^{A_{n-1}} \log h_t^s(z_n|Z^{n-1})$ for some sequence (A_n) such that $A_n \in U_n$. However, for the purposes of this book, we use the information defined by Eq. (3.4).)

3.2 Properties of Information

We now show some properties of the information measures defined above that will prove to be useful in the following chapters.

Theorem 3.1. *Let S be a parameter space. Then, for any $s \in S$ and for each $n \geq 0$, we have*

$$I_n(*;s) \geq 0 \quad \text{a.e.}$$

and

$$\overline{I}_n(*;s) \geq 0$$

with equality if and only if $f_s(z_n|Z^{n-1}) = f_*(z_n|Z^{n-1})$ a.e.

Proof.

$$I_n(*;s) = -E_*^{U_{n-1}} \log h_*^s(z_n|Z^{n-1})$$

Using the inequality

$$\log a \le a - 1; \log a = a - 1 \quad \text{if and only if} \quad a = 1 \tag{3.5}$$

we get

$$I_n(*;s) \ge 1 - E_*^{U_{N-1}} h_*^s(z_n|Z^{n-1}) = 0 \quad \text{a.e.} \tag{3.6}$$

where the second equality follows from Eq. (2.22). To show that equality holds only if $f_s(z_n|Z^{n-1}) = f_*(z_n|Z^{n-1})$ a.e. (sufficiency is trivial), suppose that

$$I_n(*;s) = 1 - E_*^{U_{N-1}} h_*^s(z_n|Z^{n-1}) \quad \text{a.e.}$$

i.e.,

$$E_*^{U_{n-1}}\{h_*^s(z_n|Z^{n-1}) - \log h_*^s(z_n|Z^{n-1}) - 1\} = 0 \quad \text{a.e.}$$

By Eq. (2.1), we then have

$$\int [h_*^s(z_n|Z^{n-1}) - \log h_*^s(z_n|Z^{n-1}) - 1] dP_* - 0 \tag{3.7}$$

Equations (3.7) and (3.5) together yield

$$h_*^s(z_n|Z^{n-1}) = 1 \quad \text{a.e.}$$

or

$$f_s(z_n|Z^{n-1}) = f_*(z_n|Z^{n-1}) \quad \text{a.e.} \tag{3.8}$$

Hence, equality in Eq. (3.6) holds if and only if Eq. (3.8) holds. Similarly, since $I_n(*;s) \ge 0$, we have

$$\bar{I}_n(*;s) = E_* I_n(*;s) \ge 0$$

with equality if and only if $I_n(*;s) = 0$ a.e. which, as shown above, occurs if and only if $f_s(z_n|Z^{n-1}) = f_*(z_n|Z^{n-1})$ a.e. □

Corollary 3.1. *Suppose that $r \in S$ is the true parameter. Then for any $t \in S, I_n(s;t)$ and $\overline{I}_n(s;t)$ are maximized on S at $s = r$. This maximum is unique unless for some $s \in S$*

$$f_s(z_n|Z^{n-1}) = f_r(z_n|Z^{n-1}) \quad \text{a.e.}$$

Proof. By Theorem 3.1, we have

$$I_n(r;t) - I_n(s;t) = I_n(r;s) \geq 0 \quad \text{a.e.}$$

and

$$\overline{I}_n(r;t) - \overline{I}_n(s;t) = \overline{I}_n(r;s) \geq 0$$

with equality if and only if $f_s(z_n|Z^{n-1}) = f_r(z_n|Z^{n-1})$ a.e. The assertion follows. □

Theorem 3.2. *The sequence $(|\overline{I}_n(s;t)|); s,t \in S$ is a sequence of pseudo-metrics on S. It is a sequence of metrics on S if and only if $\overline{I}_n(s;t) = 0$ implies $s = t$. The sequence $(|I_n(s;t)|); s,t \in S$ is a stochastic sequence of pseudo-metrics on S. It is a stochastic sequence of metrics if and only if $I_n(s;t) = 0$ implies $s = t$.*

Proof. To prove that $\overline{I}_n(s;t)$ is a pseudo-metric on S for each n, we have to show (see Section 2.6) that for each n, it satisfies the following conditions:

(i) $|\overline{I}_n(s;s)| = 0$ for any $s \in S$,
(ii) $|\overline{I}_n(s;t)| = |\overline{I}_n(t;s)|$ for any $s,t \in S$,
(iii) $|\overline{I}_n(s;t)| \leq |\overline{I}_n(s;r)| + |\overline{I}_n(r;t)|$ for any $s,t,r \in S$.

We have

$$h_s^s(z_n|Z^{n-1}) = 1$$

Hence,

$$\log h_s^s(z_n|Z^{n-1}) = 0$$

Then also

$$E \log h_s^s(z_n|Z^{n-1}) = 0$$

and (i) follows. Also,

$$\overline{I}(s;t) = -|\overline{I}(t;s)|$$

and (ii) follows.

Condition (iii) is proved as follows:

$$\begin{aligned}
&|\overline{I}_n(s;r)| + |\overline{I}_n(r;t)| \\
&= |E_* \log h_r^s(z_n|Z^{n-1})| + |E_* \log h_t^r(z_n|Z^{n-1})| \\
&= |E_* \log f_s(z_n|Z^{n-1})| - |E_* \log f_r(z_n|Z^{n-1})| \\
&+ |E_* \log f_r(z_n|Z^{n-1})| - |E_* \log f_t(z_n|Z^{n-1})| \\
&\geq |E_* \log f_s(z_n|Z^{n-1})| - |E_* \log f_t(z_n|Z^{n-1})| \\
&= |E_* \log h_t^s(z_n|Z^{n-1})| = |\overline{I}_n(s;t)|
\end{aligned}$$

if in addition to (i), (ii), and (iii), $|\overline{I}_n(s;t)|$ satisfies

(iv) $|\overline{I}_n(s;t)| = 0;\ s,t \in S$ implies $s = t$

then $|\overline{I}_n(s;t)|$ is a metric on S. The assertion follows for $|\overline{I}_n(s;t)|$. The result for $|I_n(s;t)|$ is obtained by showing that conditions (i)–(iv) above hold a.e., replacing E_* by $\mathrm{E}_*^{U_{n-1}}$ and following the same steps. □

Theorem 3.3. *For any $t, r \in S$ and for each $n \geq 0$, the sequences $(|\overline{I}_n(*;t)|)$ and $(|I_n(*;t)|)$ satisfy the properties* (i)–(iii) *above. They satisfy the property* (iv) *if and only if $f_t(z_n|Z^{n-1}) = f_*(z_n|Z^{n-1})$ a.e. implies $t = *$.*

Proof. The proof of properties (i)–(iii) is obtained precisely as in the proof of Theorem 3.2. (iv) is satisfied if and only if $f_t(z_n|Z^{n-1}) = f_*(z_n|Z^{n-1})$ a.e. implies $t = *$ by Theorem 3.1.

The variables $(|\overline{I}_n|(*;t))$ and $(I_n(*;t))$, $t \in S$, are then distance measures from the true parameters $*$ to points in the parameter set S. They can be regarded as extensions of the metrics $(|\overline{I}_n|(s;t))$ and $(I_n(s;t))$ on S to the set $T = (* \cup S)$ □

Corollary 3.2. *Let $s, t \in S$ be any pair of parameters in the parameter space S. Then s is closer to the true parameter $*$ than t in the metric $|I_n(s;t)|$ if and only if $I_n(s;t) > 0$ a.e. and in the metric $|\overline{I}_n(s;t)|$ if and only if $\overline{I}_n(s;t) > 0$.*

Proof. s is closer to the true parameter than t in the metric $|\overline{I}_n(s;t)|$ if and only if

$$|\overline{I}_n(*;s)| < |\overline{I}_n(*;t)| \quad \text{a.e.}$$

But by Theorem 3.1,

$$|\overline{I}_n(*;s)| <= \overline{I}_n(*;t) \text{ a.e.} \quad \text{for any s} \in \text{S}$$

Hence, s is closer to the true parameter than t if and only if

$$I_n(*;s) < I_n(*;t) \quad \text{a.e.}$$

or

$$I_n(*;t) - I_n(*;s) = I_n(s;t) > 0 \quad \text{a.e.}$$

To show that s is closer to the parameter than t in the metric $|\overline{I}_n(s;t)|$ an identical procedure can be followed using $\overline{I}_n(s;t)$ instead of $I_n(s;t)$. □

We shall use the notation

$$\delta_n(s;t) \equiv |I_n(s;t)|$$

and

$$d_n\,(s;t) \equiv \left|\overline{I}_n\,(s;t)\right|$$

Then we have sequences of metric spaces

$$(S,\delta_n); (S,d_n)$$

where S is the parameter set. Note that while $I_n(s;t)$ and $\delta_n(s;t)$ are U_{n-1}-measurable random variables, $\overline{I}_n(s;t)$ and $d_n(s;t)$ are not random variables. We shall see that $I_n(s;t)$ and $\delta_n(s;t)$ are useful for purposes of analysis. The metric $d_n(s;t)$ will prove particularly useful when it is constant in time, as it will prove to be the case for ergodic observation sequences. The parameter metric space can then be denoted (S,d).

3.3 Comparison with Other Information Measures

Attempts by statisticians and engineers to assign quantitative measures to the intuitive notion of information have resulted over the years in many different definitions of information. Information measures can, in essence, be classified into two different categories. One is characterized by the Shannon entropy, which has proved to be useful in communication and source coding theory, sometimes termed information theory. The other is characterized by Fisher's and Kullback's information measures, which have been more popular in statistical circles. Our information measures fall in the second category. It seems that different permutations of Fisher's or Kullback's information measures result from different interpretations of a given set of data which, in turn, reflect the intended application. Our version of information seems to be the most general, since, unlike the other definition, it does not assume that the true parameter belongs to the parameter set under consideration. However, special care must be taken in evaluating the advantages of one definition of information over another.

The information measures defined in this chapter prove to be very useful in the analysis of the asymptotic behavior of parameter estimates. They provide insight into the convergence of the estimates in the presence and absence of the true parameter. However, they can only be computed if the true parameter is known. Nevertheless, their application is not limited to analysis, as will be evident in Chapter 5, where we consider several model selection problems. On the other hand, several other information measures which are useful in given applications, such as signal detection, do not possess properties which are useful for analytical purposes, such as the metric property. In the remainder of this chapter, we discuss a few information measures, common in the information-theoretic and statistical literature, and relate them to the information measures defined here.

3.3.1 *Kullback's information, the divergence, the Bhattacharyya distance and the ambiguity function*

Kullback (1959) defined the mean information for the discrimination in favor of one hypothesis, H_1, against another, H_2, given an observation x as

$$I^K(1;2) = \int \log \frac{f_1(x)}{f_2(x)} d\mu_1(x)$$

where μ_1 is a probability measure corresponding to $H_1 \cdot f_1$ is the density of μ_1 with respect to some measure λ and f_2 is the density with respect to λ of μ_2, a probability measure corresponding to H_2. The divergence between H_1 and H_2, first introduced by Jeffreys (1946) and employed by Kullback (1959) is defined as

$$\begin{aligned} J(1;2) &= I^k(1;2) + I^k(2;1) \\ &= \int [f_1(x) - f_2(x)] \log \frac{f_1(x)}{f_2(x)} d\lambda(x) \\ &= \int \log \frac{f_1(x)}{f_2(x)} d\mu_1(x) - \int \log \frac{f_1(x)}{f_2(x)} d\mu_2(x) \end{aligned}$$

In contrast, $I(1;2)$, defined by Eq. (3.3), would be written as

$$I(1;2) = \int \log \frac{f_1(x)}{f_2(x)} \mathrm{d}\mu(x)$$

where $\mu(x)$, the correct probability measure, may be different from both $\mu_1(x)$ and $\mu_2(x)$.

The Bhattacharyya distance (Bhattacharyya, 1943) between two densities $f_1(x)$ and $f_2(x)$ of an observation x is represented by

$$B = -\ln \int [f_1(x) f_2(x)]^{1/2} d\lambda$$

where λ is the Lebesgue measure on the space of x. Properties of the Bhattacharyya distance and the divergence were studied and compared by Kailath (1967), and they were found to be particularly suitable for signal detection in communication. However, Kullback's information, the divergence and the Bhattacharyya distance do not satisfy the triangle inequality and thus fail as metrics on the parameter (or hypothesis) space. In contrast, the metric property of the information measures introduced in Section 3.1 follows from the consistent use of the probability measure throughout, whereas Kullback's information, the divergence and the Bhattacharyya distance are defined using different measures.

The ambiguity in an observation x between a parameter s and the true parameter $*$ is defined as

$$\gamma_s = E_* \log f_s(x)$$

The ambiguity function γ_s has been found to be useful in the analysis of error in radar applications (Woodward, 1953). In fact,

$$I(s;t) = \gamma_t - \gamma_s$$

Hence, the information between two parameters as defined in this book is the difference between their ambiguities.

3.3.2 *Fisher's information*

We shall now show that the information measure introduced in Section 3.1 is related to Fisher's information measure (Fisher, 1956; Savage, 1954). We follow a similar comparison between Kullback's and Fisher's information measures (Kullback, 1959). However, in order to relate measures of the same quantity, we define Fisher's information in a single observation z_n.

Let $S \in R^K$ be the parameter space. Suppose that for any $s \in S$, the following regularity conditions (Cramer, 1946; Gurland, 1954) hold for all $i,\ j = 1, \ldots, k$:

(1) $$\frac{\partial \log f_s(z_n|Z^{n-1})}{\partial s^i} < F_1(Z^n); \quad \frac{\partial^2 \log f_s(z_n|Z^{n-1})}{\partial s^i \partial s^j} < F_2(Z^n) \quad \text{a.e.}$$

where the partial derivatives are assumed to exist and $F_1(z^n)$ and $F_2(z^n)$ are integrable random variables.

(2) $$\int \frac{\partial f_s(z_n|Z^{n-1})}{\partial s^i} \mathrm{dP}_* = 0; \quad \int \frac{\partial^2 f_s(z_n|Z^{n-1})}{\partial s^i \partial s^j} \mathrm{dP}_* = 0$$

We define Fisher's information in a single observation at a parameter point s as a matrix $\overline{I}_n^F$, whose elements are

$$\overline{I}_{i,J,n}^s \equiv E_* \left\{ \left(\frac{1}{f_s(z_n|Z^{n-1})} \frac{\partial f_s(z_n|Z^{n-1})}{\partial s^i} \right) \times \left(\frac{1}{f_s(z_n|Z^{n-1})} \frac{\partial f_s(z_n|Z^{n-1})}{\partial s^j} \right) \right\} \tag{3.9}$$

Consider a point $s \in S$ and a close point $s + \Delta s \in S$. Using Taylor's expansion to second order (e.g., Struik, 1969), we have

$$\log f_{s+\Delta s}(z_n|Z^{n-1}) - \log f_s(z_n|Z^{n-1})$$
$$= \sum_{i=1}^{k} \Delta s^i \tfrac{\partial \log f_s(z_n|Z^{n-1})}{\partial s^i} + \frac{1}{2}\sum_{i=1}^{k}\sum_{j=1}^{k} \Delta s^i \Delta s^j \frac{\partial^2 \log f_s(z_n|Z^{n-1})}{\partial s^i \partial s^j}$$

But

$$\frac{\partial \log f_s(z_n|Z^{n-1})}{\partial s^i} = \frac{1}{f_s(z_n|Z^{n-1})}\frac{\partial f_s(z_n|Z^{n-1})}{\partial s^i};$$
$$\frac{\partial^2 \log f_s(z_n|Z^{n-1})}{\partial s^i \partial s^j} = \frac{1}{f_s(z_n|Z^{n-1})}\frac{\partial^2 f_s(z_n|Z^{n-1})}{\partial s^i \partial s^j}$$
$$- \frac{1}{f_s^2(z_n|Z^{n-1})}\frac{\partial f_s(z_n|Z^{n-1})}{\partial s^i}\frac{\partial f_s(z_n|Z^{n-1})}{\partial s^j}$$

The information in z_n favoring s against a closed point $s + \Delta s$ as defined by Eq. (3.4) is

$$\overline{I}_n(s; +\Delta s) = \int \log \frac{f_s(z_n|Z^{n-1})}{f_{s+\Delta s}(z_n|Z^{n-1})} \mathrm{dP}_*$$
$$= -\int \sum_{i=1}^{k} \Delta s^i \frac{\partial \log f_s(z_n|Z^{n-1})}{\partial s^i} \mathrm{dP}_*$$
$$- \frac{1}{2}\int \sum_{i=1}^{k}\sum_{j=1}^{k} \Delta s^i \Delta s^j \frac{\partial^2 \log f_s(z_n|Z^{n-1})}{\partial s^i \partial s^j} \mathrm{dP}_*$$
$$= -\sum_{i=1}^{k} \Delta s^i \int \frac{\partial f_s(z_n|Z^{n-1})}{\partial s^i} \mathrm{dP}_*$$
$$- \frac{1}{2}\sum_{i=1}^{k}\sum_{j=1}^{k} \Delta s^i \Delta s^j \int \left[\frac{\partial^2 \log f_s(z_n|Z^{n-1})}{\partial s^i \partial s^j}\right.$$
$$\left. - \frac{1}{f_s(z|Z^{n-1})}\frac{\partial f_s(z|Z^{n-1})}{\partial s^i}\frac{\partial f_s(z|Z^{n-1})}{\partial s^j}\right] \mathrm{dP}_*$$
$$= \frac{1}{2}\sum_{i=1}^{k}\sum_{j=1}^{k} \Delta s^i \Delta s^j \overline{I}_{i,j}^F(s)$$

where the last equality is obtained by the regularity condition (2) above. Hence, the information in a single observation is related to Fisher's information in a single observation by

$$\overline{I}_n(s; s + \Delta s) = \frac{1}{2}\Delta s^T \overline{I}_n^F(s)\Delta s$$

Defining similarly the conditional Fisher Information in a single observation as a matrix $I_n^F(s)$ whose elements are

$$I_{i,j,n}^F(s) \equiv E_*^{U_{n-1}}\left\{\left(\frac{1}{f_s(z_n|Z^{n-1})}\frac{\partial f_s(z_n|Z^{n-1})}{\partial s^i}\right)\right.$$
$$\left.\times\left(\frac{1}{f_s(z_n|Z^{n-1})}\frac{\partial f_s(z_n|Z^{n-1})}{\partial s^i}\right)\right\}$$

we get, using a similar procedure,

$$I_n(s; s + \Delta s) = \frac{1}{2}\Delta s^T I_n^F(s)\Delta s$$

3.3.3 *Self-information and entropy*

To complete this discussion, placing the information measures motivated and defined in this chapter with respect to other measures found in the literature, we mention two other measures which are common in information theory, namely, the self-information and the entropy (e.g., Fano, 1961; Gallager, 1968). The definitions of these measures are based on the Bayesian assumption (see Chapter 2).

Consider a parameter set S. The self-information in the measurement Z^n about a parameter $s \in S$ is defined as

$$I_n^s(s) \equiv -\log f^b(s|Z^n) \tag{3.10}$$

A comparative measure of information can then be obtained by taking the difference of the self-information corresponding to the two parameters:

$$\Delta I_n^s(s:t) \equiv I_n^s(s) - I^s(t) = -\log\frac{f^b(s|Z^n)}{f^b(t|Z^n)}$$

The self-information between s and t in a single observation z_n can be obtained using Eqs. (2.7) and (2.20) as

$$\Delta I_n^s(s;t) - \Delta_{n-1}^s(s;t) = -\log \frac{f_s(z_n|Z^{n-1})}{f_t(z_n|Z^{n-1})} = -\log h_t^s(z_n|Z^{n-1}) \tag{3.11}$$

Taking the expectation and the conditional expectation of Eq. (3.11) with respect to the true measure, one gets

$$E_*\{\Delta I_n^s(s;t) - \Delta I_{n-1}^s(s;t)\} = -\bar{I}_n(s;t) \tag{3.12}$$

and

$$E_*^{U_{n-1}}\{\Delta I_n^s(s;t) - \Delta I_{n-1}^s(s;t)\} = -I_n(s;t) \tag{3.13}$$

Hence, the mean and the conditional mean values of the self-information difference in a single observation are the negative values of the information measures defined in Section 3.1. (The sign is, of course, of no significance since the self-information defined by Eq. (3.6) is, in fact, a lack of information and would become positive information in the sense meant in this chapter by inverting the sign.)

Note that in Eq. (3.9), the expectation is taken with respect to the correct probability measure P_*, independently of whether the correct parameter ever belongs to the set S. If, on the other hand, one makes the assumption that the true parameter belongs to a finite set, say, $\{s_j; j \in k \equiv (0, \ldots, p)\}$, and takes a conditional expectation given Z^n of Eq. (3.6), then one gets

$$E^{b^{U_n}} I_n^s(s;) = \sum_{j=0}^{p} f^b(s_j|Z^n) \log f^b(s_j|Z^n) \equiv H(Z^n) \tag{3.14}$$

Equation (3.14) is the entropy in Z^n. Note that the entropy differs from the other information measures considered in this section in the sense that it is not a comparative measure between parameters. It does provide some measure of the average information gained from the observation with respect to the *a priori* assumptions (Berger, 1971).

Chapter 4

Convergence of Maximum Likelihood and Bayesian Estimates on Finite Sets of Parameters

Before analyzing the selection of a linear control model from a finite set of such models, as done in the following chapter, we address the selection from a set of more general models. The underlying selection process involves convergence of maximal likelihood and Bayesian parameter estimates over general classes of observation sequences. The convergence of the estimates follows from the convergence of the likelihood ratios over the parameter set. Consistency conditions are derived in terms of the information in the observations. The case where the true parameter is not necessarily a member of the parameter set is also considered. Rates of convergence in the mean for the maximum likelihood (ML) and the maximum *a posteriori* distribution (MAP) procedures are derived.

4.1 Convergence of Parameter Estimates

Let (z_n) be a stochastic process on a probability space (Ω, U, P_K) and let $S = K \equiv \{0, \ldots, p\}$ be a parameter set such that $\{P_j; j \in K\}$ is a family of probability measures on (Ω, U). Let (U_n) be an increasing sequence of σ-subalgebras of U generated by (Z^n) and let $P_{j,n}$ be the restriction of P_j to U_n for each $j \in K$. Consider the following condition:

(c4.1) For some $k \in K$ and for each $j \in K; j \neq k$,

$$\lim_{n\to\infty} h_k^j(Z^n) = 0 \quad \text{a.e.} \tag{4.1}$$

We next show that the convergence a.e. of the ML and MAP parameter estimation procedures and the convergence a.e. in m.s. of the LS procedure follow from condition (c4.1). Naturally, a major difficulty in proving convergence of the parameter estimates is to verify condition (c4.1). In the following chapters, we present the conditions for general classes of observation sequences under which condition (c4.1) is satisfied when k is the true parameter and extend the results to the case where the true parameter is not necessarily a member of the parameter set. The latter case is treated specifically in the following four theorems, which will prove very useful.

Theorem 4.1. *Suppose that condition* (c4.1) *is satisfied, then ML estimates on K converge a.e. to k as $n \to \infty$.*

Proof. Since the set $j \in K; j \neq k$ is finite, condition (c4.1) implies

$$\lim_{n\to\infty} \sup_j\{h_k^j(Z^n); j \in K; j \notin k\} = 0 \quad \text{a.e.}$$

Hence,

$$\lim_{n\to\infty} \sup_j\{h_k^j(Z^n); j \in K\} = h_k^k(Z^n) = 1 \quad \text{a.e.}$$

or

$$\lim_{n\to\infty} \widehat{k}(Z^n) = k \quad \text{a.e.}$$

□

Theorem 4.2. *Under condition* (c4.1), *MAP estimates on K converge a.e. to k.*

Proof. By Eq. (2.20), we have, for each $k \in K$,

$$f^b(j|Z^n) \leq \frac{f_0^b(j) f_j(z^n)}{f_0^b(k) f_k(z^n)} = \frac{f_0^b(j)}{f_0^b(k)} h_k^j(Z^n)$$

By condition (c4.1), for each $k \in K; j \neq k$, we have

$$\lim_{n\to\infty} f^b(j|Z^n) \leq \frac{f_0^b(j)}{f_0^b(k)} \lim_{n\to\infty} h_k^j(Z^n) = 0 \quad \text{a.e.}$$

implying

$$\lim_{n\to\infty} f^b(j|Z^n) = 0 \quad \text{a.e.} \quad \text{for each } j \in K; \quad j \neq k \tag{4.2a}$$

But since

$$\sum_{j=0}^{p} f^b(j|Z^n) = 1$$

we have

$$\lim_{n\to\infty} f^b(k|Z^n) = 1 \quad \text{a.e.} \tag{4.2b}$$

yielding the assertion. □

Theorem 4.3. *Suppose that a parameter vector s is assumed to belong to a finite set $s_j \in R^m; j \in K$ in the calculation of the estimates (but is not necessarily a member of the set). Suppose further that, for some $k \in K$, condition (c4.1) is satisfied. Then, the LS estimates of s on R^m converge a.e. to s_k.*

Proof. By Eqs. (2.21) and (4.2), we have

$$\lim_{n\to\infty} \widehat{s}_n = \sum_{j=0}^{p} s_j \lim_{n\to\infty} f^b(j|Z^n) = s_k \quad \text{a.e.}$$ □

Theorem 4.4. *For the situation given in Theorem 4.3, LS estimates converge to s_k in the mean square.*

Proof. We follow in part Liporace (1971), who treated the case of independent and identically distributed observations. Consider the norm

$$\begin{aligned} N_n &= E_*\{(\widehat{s}_n - s_k)^T(\widehat{s}_n - s_k)\} \\ &= E_*\left\{\sum_{j=0}^{p}(s_j - s_k)^T f^b(s_j|Z^n) \sum_{i=0}^{p}(\widehat{s}_i - s_k) f^b(s_j|Z^n)\right\} \end{aligned}$$

$$= \sum_{j=0}^{p} \sum_{i=0}^{p} (s_j - s_k)^T (s_i - s_k) E_* \{ f^b(s_j|Z^n) f^b(s_j|Z^n) \}$$

$$\leq p^2 R^2 E_* f^b(s_j|Z^n) \quad \text{for some } j \in K; j \neq k$$

where

$$R^2 = \max\{(s_j - s_k)^T (s_j - s_k); j \in K; j \neq K\}$$

since obviously

$$E_*\{f^b(s_j|Z^n)\} \geq E_*\{f(s_j|Z^n) f(s_i|Z^n)\} \quad \text{as } f(s_i|Z^n) \leq 1.$$

By Eq. (2.20), we have, for each $j \in K$,

$$f^b(s_j|Z^n) \leq \frac{f_0^b(s_j) f_j(Z^n)}{f_0^b(s_k) f_k(Z^n)} = \frac{f_0^b(s_j)}{f_0^b(s_k)} h_k^j(Z^n)$$

By condition (c4.1), we have, for each $j \in K$; $j \neq k$,

$$\lim_{n\to\infty} f^b(s_j|Z^n) \leq \frac{f_0^b(s_j)}{f_0^b(s_k)} \lim_{n\to\infty} h_k^j(Z^n) = 0 \quad \text{a.e.}$$

Hence,

$$\lim_{n\to\infty} f^b(s_j|Z^n) = 0 \quad \text{a.e.}$$

Now, since

$$f^b(s_j|Z^n) \leq 1$$

we have by the dominated convergence theorem (Theorem 2.3) that, for each $j \in k; j \neq k$,

$$\lim_{n\to\infty} E_* f^b(s_j|Z^n) = E_* \lim_{n\to\infty} f^b(s_j|Z^n) = 0$$

and thus,

$$\lim_{n\to\infty} N_n \leq p^2 R^2 \lim_{n\to\infty} f^b(s_j|Z^{n-1}) = 0$$

yielding finally

$$\lim_{n\to\infty} N_n = 0.$$

□

4.2 Consistency of the Estimates

In Chapter 3, we defined, for each pair $k, j \in K$,

$$I_n(k;j) \equiv E_*^{U_{n-1}} \log h_j^k(z_n|Z^{n-1})$$

Let us also define

$$J_n(k;j) \equiv \log h_j^k(z_n|Z^{n-1}) - I_n(k;j)$$

$J_n(k;j)$ is the error in the incremental information $I_n(k;j)$ or the information residual.

Denote

$$Y_n(k;j) \equiv \sum_{m=1}^{n} I_m(k;j)$$

and

$$V_n(k;j) \equiv \sum_{m=1}^{n} J_m(k;j)$$

Suppose that $* \in K$, i.e., that the true parameter is a member of the parameter set, and consider the following conditions:

(c4.2) For some $k \in K$ and for each $j \in K; j \neq k$,

$$\limsup V_n(k;j) > -\infty \quad \text{a.e.}$$

(c4.3) For some $k \in K$ and for each $j \in K; j \neq k$,

$$\lim_{n\to\infty} Y_n(k;j) = \infty \quad \text{a.e.}$$

Lemma 4.1. *Suppose that conditions* (c4.2) *and* (c4.3) *hold for* $k = *$. *Then, for each* $j \in k; j \neq *$, *we have*

$$\lim_{n\to\infty} h_*^j(z^n) = 0 \quad \text{a.e.}$$

Proof. We have noted (see Section 2.4) that for each $j \in K$, the sequence $(h_*^j, (z^n))$ is a U_n- martingale according to the measure P_*.

Furthermore,

$$E_* h_*^j(Z^n) = 1$$

It follows from the martingale convergence theorem (Theorem 2.4) that the sequence $(h_*^j, (z^n))$ converges to a finite limit. Thus, the sequence $(\log h_*^j, (z^n))$ converges to some $a < \infty$. We have

$$\log h_j^*, (z^n) = Y_n(*; j) + V_n(*; j) \tag{4.3}$$

Suppose that

$$\lim_{n\to\infty} \log h_*^j(Z^n) = a > -\infty, \quad \text{a.e.}$$

or

$$\lim_{n\to\infty} \log h_j^*(Z^n) = a < \infty \quad \text{a.e.}$$

Then, by condition (c4.3) and by Eq. (4.3), we have

$$\lim_{n\to\infty} V_n(*; j) = -\infty \quad \text{a.e.}$$

contradicting condition (c4.2). Hence, we have

$$\lim_{n\to\infty} \log h_j^*(Z^n) = \infty \quad \text{a.e.}$$

or

$$\lim_{n\to\infty} \log h_*^j(Z^n) = -\infty \quad \text{a.e.}$$

yielding

$$\lim_{n\to\infty} h_*^j(Z^n) = 0 \quad \text{a.e.}$$

□

Theorem 4.5. *Suppose that some $k \in K$ is the true parameter. Then under conditions* (c4.2) *and* (c4.3), *the ML and MAP estimates are consistent a.e. and the LS estimates are consistent a.e. and in the mean square.*

Proof. The assertion follows directly from Lemma 4.1 and from Theorems 4.1–4.4. □

Consider the following condition:

(c4.4) For some $k \in K$ and for each $j \in K$; $j \neq k$, there exists some $\varepsilon_j > 0$ and a subsequence (n_{i_j}) of n such that

$$I_{n_{i_j}}(k;j) \geq \varepsilon_j \quad \text{a.e. for all } n_{i_j}$$

Theorem 4.6. *Suppose that some $k \in K$ is the true parameter. Then, under conditions* (c4.2) *and* (c4.4), *the ML and MAP estimates are consistent a.e. and in the mean square.*

Proof. By Theorem 3.1, we have

$$I_n(k;j) \geq 0 \quad \text{a.e. for all } n \geq 0$$

Thus, condition (c4.4) implies condition (c4.3). The assertion follows from Theorem 4.5. □

In the following chapters, we shall see certain important cases to which the information condition (c4.3) applies. We now examine condition (c4.2). We have noted that for each pair $j, k \in K$, the sequence $(J_n(k;j))$ is a martingale difference sequence according to the true measure P_*. The following special case is of particular interest.

Lemma 4.2. *For any pair $j, k \in K$, let $(J_n(k;j))$ be an ergodic sequence. Then*

$$\limsup V_n(k;j) = \infty \quad \text{a.e.}$$

Proof. We have by Eq. (2.22), for each $\omega \in \Omega$,

$$V_n(k;j,\omega) = J_1(k;j,\omega) + V_{n-1}(k;j,T\omega)$$

where T is a measure-preserving transformation. It follows that the event

$$\{\lim_{x\to\infty} \sup V_n(k;j) < \infty\}$$

is invariant. Thus, either

$$P\{\lim_{x\to\infty} \sup V_n(k;j) < \infty\} = 0$$

or

$$P\{\lim_{x\to\infty} \sup V_n(k;j) < \infty\} = 1$$

Obviously, we have that if

$$\lim_{x\to\infty} \sup V_n(k;j) < \infty$$

then

$$\lim_{x\to\infty} \sup \frac{V_n(k;j)}{\sqrt{n}} < \infty$$

But, by Theorem 2.6,

$$P\left\{\lim_{x\to\infty} \sup \frac{V_n(k;j)}{\sqrt{n}} < \infty\right\} < 1$$

Hence,

$$P\{\lim\sup V_n(k;j) \ < \infty\}< 1$$

yielding

$$P\{\lim\sup V_n(k;j) < \infty\} = 0$$

Thus,

$$\lim\sup V_n(k;j) = \infty \quad \text{a.e.}$$

□

Example 4.1. Let (x_n) be a sequence of independent identically distributed observations. Suppose that each x_n is distributed according to the density

$$f(x_n) = \frac{1}{\sqrt{2\pi\sigma^2}} e^{-\frac{x_n^2}{2\sigma^2}}$$

Let the covariance σ^2 be given on a set $\{\sigma_i, i = 1, 2\}$ and suppose that $\sigma^2 = \sigma_1^2$, i.e., 1 is the true parameter. As in Example 3.1, we have, for all $n \geq 0$,

$$I_n(1;2) = \frac{1}{2}\log\frac{\sigma_2^2}{\sigma_1^2} + \frac{\sigma_1^2}{2}\left(\frac{1}{\sigma_2^2} - \frac{1}{\sigma_1^2}\right)$$

and

$$J_n(1;2) = \frac{1}{2}x_n^2\left(\frac{1}{\sigma_2^2} - \frac{1}{\sigma_1^2}\right) - \frac{\sigma_1^2}{2}\left(\frac{1}{\sigma_2^2} - \frac{1}{\sigma_1^2}\right)$$

Since (x_n) is an ergodic sequence, so is $j_n(1;2)$. Thus, by Lemma 4.2, condition (c4.2) is satisfied. It follows from Eq. (3.5) that if $\sigma_1 \neq \sigma_2$, then $I_n(1;2) \neq 0$ and then, by Theorem 3.1, we have

$$I_n(1;2) = I(1;2) > 0 \quad \text{for all } n \geq 0$$

Thus, condition (c4.3) is satisfied for $k = 1$. Hence, by Theorems 4.1–4.4, the ML and MAP estimates of σ will converge a.e. and the LS estimate will converge a.e. and in the mean squares to σ_1.

The following general result provides a sufficient condition satisfying condition (c4.2). Although it will not be used directly in the following chapters, it seems to have useful implications (see Example 4.2).

Lemma 4.3. *Suppose that for any $j, k \in K$, we have, for any positive scalar a,*

$$E_*\{V_{n_a}(k;j) - a\} < \infty \tag{4.4}$$

where

$$n_a = \inf\{n : V_n(k;j) > a\} \tag{4.5}$$

then, for each $\omega \in \Omega$, either

$$\lim_{n\to\infty} V_n(k;j,\omega) \, \text{exists and is finite}$$

or

$$\limsup V_n(k;j,\omega) = \infty$$

Proof. Let

$$R_a(k;j) \equiv v_{n_a}(k;j) - a$$

and

$$v_n^a(k;j) \equiv v(k;j)_{\min(n,n_a)}$$

Note that

$$v_n^a(k;j) \leq a + R_a(k;j)$$

Since $v_n(k;j)$ is a U_n-martingale, so is $v_n^a(k;j)$. Obviously, we have

$$E_*\{v_n^{a^+}(k;j)\} \leq a + E_* R_a(k;j)$$

Hence, under Eq. (4.4),

$$E_*\{v_n^{a^+}(k;j)\} \leq \infty$$

It follows from Theorem 2.4 that the sequence $v_n^a(k;j)$ converges to a finite limit. Let

$$A_a \equiv \{\omega \in \Omega; \sup V_n(k;j,\omega) < a\}$$

and

$$A \equiv \cup_{a=1}^{\infty} A_a$$

If $\omega \in A$ for some a say, a_0, then

$$v_n(k;j,\omega) = v_n^{a_0}(k;j,\omega) \quad \text{for all } n$$

and then

$$\lim_{n\to\infty} v_n(k;j,\omega) \equiv \lim_{n\to\infty} v_n^{a_0}(k;j,\omega) \text{ is finite.}$$

If $\omega \notin A$, then

$$\limsup v_n(k;j,\omega) = \infty. \qquad \square$$

Example 4.2. Let (x_n) be a sequence of real-valued random variables taking values in the interval [0, 3]. Suppose that the sequence (x_n) is not necessarily independent or identically distributed. Consider two hypotheses (or two parameters) 1 and 2, according to which (x_n) i.i.d. with probability densities

$$f_1(x_n) = \begin{cases} \frac{1}{2} & 0 \leq x_n \leq 1 \\ \frac{1}{4} & 1 \leq x_n \leq 1 \\ 0 & \text{elsewhere} \end{cases}$$

and

$$f_2(x_n) = \begin{cases} \frac{1}{4} & 0 \leq x_n \leq 2 \\ \frac{1}{2} & 2 \leq x_n \leq 3 \\ 0 & \text{elsewhere} \end{cases}$$

It is easy to see that

$$J_n(1;2) \leq 2\log 2 \quad \text{for all } n$$

independently of the actual values of the sequence (x_n) might take in the interval [0, 3].

Now, since for any $a > 0$,

$$V_{n_a}(1;2) = V_{n_a-1} + J_{n_a}(1;2) \leq a + J_{n_a}(1;2)$$

we have

$$E_*\{V_{n_a}(1;2) - a\} \leq E_* J_{n_a}(1;2) \leq 2\log 2$$

for all n. Hence, Eq. (4.4) holds. It follows from Lemma 4.2 that condition (c4.2) is satisfied for this case independently of the actual probability measure generating the sequence (x_n).

4.3 Convergence in the Absence of the True Parameter

Consider the probability and parameter spaces given in Section 4.1. While the absolute continuity of the restrictions $P_{1,n}$ and $P_{2,n}$ of two measures P_1 and P_2 to the σ-subalgebra U_n of U is possible to verify in practical situations (it follows, e.g., from the absolute continuity of the corresponding conditional densities $f_{1,n}(z_n|Z^{n-1})$ and $f_{2,n}(z_n|Z^{n-1})$ for each n), the absolute continuity of P_1 and P_2 does not follow and is, in general, more difficult to verify. The following results are nevertheless interesting from a theoretical viewpoint.

Theorem 4.7. *Let conditions* (c4.2) *and* (c4.3) *hold for some parameter $k \in K$. Furthermore, suppose that the true measure P_* is absolutely continuous with respect to the measure P_k. Then, for each $j \in k; j \neq k$, we have*

$$\lim_{n\to\infty} h_k^j(z^n) = 0$$

and, consequently, the parameter estimates will converge to the parameter k in the sense specified in Theorems 4.1–4.4.

Proof. Since the sequence $(h_k^j(Z^n))$ is a (U_n, P_k)-martingale and since

$$E_k h_k^j(Z^n) = 1$$

it follows from Theorem 2.4 that $(h_k^j(Z^n))$ converges a.e. P_k to a finite random variable. Since P_* is absolutely continuous with respect to P_k, then $(h_k^j(Z^n))$ converges to a finite random variable a.e. P_k.

The remainder of the proof is identical to the proof of Lemma 4.1, and the convergence of the estimates follows from the convergence of the likelihood ratios by Theorems 4.1–4.4. □

In the following chapter, we treat the case where the true parameter is not necessarily a member of the parameter set for a model of particular interest, namely, linear dynamical system. We shall not, however, investigate the absolute continuity of the probability measures P_* and P_k, but rather use simpler arguments, enabled by the particular problem under consideration.

Condition (c2.1) requires that for any parameter $k \in K$, the restriction $P_{j,n}$ of the measure $P_j; j \in K, j \neq k$ be absolutely continuous with respect to the restriction $P_{k,n}$ of the measure P_k. An interesting observation is presented by the following theorem.

Theorem 4.8. *Suppose that condition* (c4.1) *holds for the parameter $k \in K$. Then the measures $P_j, j \in K; j \neq k$ are singular with respect to the measure P_*.*

Proof. For each $j \in K$, the likelihood ratio sequence $\left(\frac{f_k(Z^n)}{f_j(Z^n)}\right)$ is a martingale according to the measure P_j (Doob, 1953, p. 93). In addition, we have

$$E_j\left(\frac{f_k(Z^n)}{f_j(Z^n)}\right) = 1 \tag{4.6}$$

By the martingale convergence theorem, we then have

$$\lim_{n\to\infty}\left(\frac{f_k(Z^n)}{f_j(Z^n)}\right) = \text{finite r.v.} \quad \text{a.e. } P_j$$

(where r.v. denotes random variable).

But under condition (c4.1),

$$\lim_{n\to\infty}\left(\frac{f_k(Z^n)}{f_j(Z^n)}\right) = \infty \quad \text{a.e. } P_*$$

So, we have

$$P_j\left\{\lim_{n\to\infty}\left(\frac{f_k(Z^n)}{f_j(Z^n)}\right) = \text{finite r.v.}\right\} = 1$$

and also,

$$P_* \left\{ \lim_{n\to\infty} \left(\frac{f_k(Z^n)}{f_j(Z^n)} \right) = \text{finite r.v.} \right\} = 0$$

Hence, under condition (c4.1), the measures $P_j; j \in K; j \neq k$ are singular with respect to the measure P_*. □

4.4 L_1 Convergence

The L_1 convergence of the likelihood and *a posteriori* probability ratios follows directly if condition (c4.1) holds. We show that under a certain condition on the information in the observation, the convergence rates are bounded by exponentials of the number of samples. The true parameter is not assumed to belong to the parameter set. These results provide performance measures for the ML and MAP estimation methods. In the following chapters, we show that bounds of the L_1 convergence rates can be computed in common situations for linear systems.

Theorem 4.9. *Suppose that condition* (c4.1) *holds for some* $k \in K$. *Then, for each* $j \in K; j \neq k$, *we have*

$$\lim_{n\to\infty} E_* h_j^k(z^n) = \infty$$

and

$$\lim_{n\to\infty} E_* \frac{f^b(k|z^n)}{f^b(j|z^n)} = \infty$$

Proof. We have by condition (c4.1), for each $j \in K; j \neq k$,

$$\lim_{n\to\infty} h_j^k(Z^n) = \infty \quad \text{a.e.}$$

and by condition (c4.2),

$$\lim_{n\to\infty} E_* \frac{f^b(k|z^n)}{f^b(j|z^n)} = \infty \quad \text{a.e.}$$

Since both sequences are non-negative, we have, by Fatou's lemma (Theorem 2.2),

$$\lim_{n\to\infty} E_* h_j^k(Z^n) \geq \lim_{n\to\infty} \inf E_* h_j^k(Z^n) \geq E_* \lim_{n\to\infty} \inf h_j^k(Z^n) = \infty$$

and

$$\lim_{n\to\infty} E_* \frac{f^b(k|z^n)}{f^b(j|z^n)} \geq \lim_{n\to\infty} \inf E_* \frac{f^b(k|z^n)}{f^b(j|z^n)} \geq E_* \lim_{n\to\infty} \inf \frac{f^b(k|z^n)}{f^b(j|z^n)} = \infty$$

□

Now, consider the following condition:

(c4.5) There exists a parameter $k \in K$ such that for each $j \in K$; $j \neq k$, there exists a positive scalar α_j such that

$$\overline{I}_n(k;j) \geq \alpha_j \quad \text{for all } n \geq N_j \tag{4.7}$$

Theorem 4.10. *Under condition* (c4.5), *there exists some positive integer N such that for each $j \in K; j \neq k$, the sequences $(h_j^k(z^n))$ and $\left(\frac{f^b(k|z^n)}{f^b(j|z^n)}\right)$ diverge in L_1 at rates no slower than exponential for all $n \geq N$.*

Proof.

$$\overline{I}_n(k;j) = E_* \log \frac{f_k(z_n|Z^{n-1})}{f_j(z_n|Z^{n-1})} = E_* \log \frac{f_k(Z^n)}{f_j(Z^n)} - E_* \log \frac{f_k(Z^n - 1)}{f_j(Z^n - 1)}$$

By condition (c4.5), we then have

$$E_* \log \frac{f_k(Z^n)}{f_j(Z^n)} - E_* \log \frac{f_k(Z^{n-1})}{f_j(Z^{n-1})} \geq \alpha_j$$

yielding

$$E_* \log \frac{f_k(Z^n)}{f_j(Z^n)} \geq a_j + (n - N_j)\alpha_j \quad \text{for all } n \geq N_j \tag{4.8}$$

where

$$a_j \equiv E_* \log \frac{f_k(z_{N_j}|Z^{N_j-1})}{f_j(z_{N_j}|Z^{N_j-1})} = \overline{I}_{N_j}(k;j) \geq \alpha_j \tag{4.9}$$

Since log (*) is a concave function, we have, by Jensen's inequality (Theorem 2.1),

$$\log E_* \frac{f_k(Z^n)}{f_j(Z^n)} \geq E_* \log \frac{f_k(Z)}{f_j(Z)} \tag{4.10}$$

Equations (4.8)–(4.10) give

$$E_* h_j^k(Z^n) = E_* \frac{f_k(Z^n)}{f_j(Z^n)} \geq e^{a_j} e^{(n-N_j)\alpha_j} \geq e^{(n-N_j+1)\alpha_j} \tag{4.11}$$

for all $n \geq N_j$ for each $j \in K; j \neq k$.

Hence, for each $j \in K; j \neq k$, the likelihood ratio $h_j^k(z^n)$ tends in the mean to infinity faster than an exponential with a rate of α_j.

By Eq. (2.20), we have, for each $j \in k$,

$$\frac{f^b(k|Z^n)}{f^b(j|Z^n)} = \frac{f^b(k)}{f^b(j)} \frac{f_k(Z^n)}{f_j(Z^n)}$$

Thus, by Eq. (4.11), for each $j \in K; j \neq k$,

$$E_* \frac{f^b(k|Z^n)}{f^b(j|Z^n)} = \frac{f^b(k)}{f^b(j)} E_* \frac{f_k(Z^n)}{f_j(Z^n)} \geq \frac{f^b(k)}{f^b(j)} e^{(n-N_j+1)\alpha_j} \quad \text{for all } n \geq N_j \tag{4.12}$$

Hence, for each $j \in K; j \neq k$, the *a posteriori* probability ratio $\frac{f^b(k|Z^n)}{f^b(j|Z^n)}$ tends in the mean to infinity faster than an exponential with a rate of α_j.

Finally, taking $N = \max\{N_j; j \in K; j \neq k\}$ and $\alpha = \min\{\alpha_j; j \in K; j \neq k\}$, the sequence $(h_j^k(z^n))$ and $\frac{f^b(k|Z^n)}{f^b(j|Z^n)}$ converge in L_1 to infinity faster than an exponential with a rate of α for all $n \geq N$. □

At instant n, the ML estimation method will select the parameter k if

$$\frac{f_k(z^n)}{f_j(z^n)} \geq 1 \quad \text{for all } j \in K;\ j \neq k \tag{4.13}$$

The MAP method will select k if

$$\frac{f^b(k|Z^N)}{f^b(j|Z^N)} \geq 1 \quad \text{for all } j \in K; j \neq k \tag{4.14}$$

Hence, the L_1 convergence bounds established in Theorem 4.10 provide a qualitative measure of performance for the ML and MAP estimates in terms of rates at which Eqs. (4.13) and (4.14) are attained in the mean. Of course, the bounds cannot be computed unless the true measure is known. Yet, if the true parameter can be assumed to belong to a finite set, then bounds can be computed over the set. This will be demonstrated in the following chapters, where we consider a specific type of model, namely, linear systems. Our consideration of model selection from a set that may or may not include the correct model was based on the statistical nature of the selection criterion rather than the mathematical nature of the models.

Chapter 5

Stationary Neural Circuit Selection from a Finite Set

In this chapter, we restrict our attention to finite sets of neural circuits, one selected at any given time for best performance. The selection is made in terms of the second-order statistics associated with the models in the model set. If the true best circuit is included in the set, it will be identified under a verifiable uniqueness condition. If the true model is not included in the model set, then the identification procedures converge to a model in the model set that is closest to the true model in the information metric sense. We show that, under a simple uniqueness condition, the sequences of likelihood and *a posteriori* probability ratios are bounded in L_1 by simple exponentials. If the true neural circuit belongs to the given model set, then the bounds can be easily computed using the *a priori* data. The L_1 convergence results provide performance measures for the ML and MAP identification methods. Finally, the analysis is extended to other modelling systems. Methods are suggested for selecting a reduced-order model to represent a high-order system and for selecting a circuit from a set, which is the true model, or an appropriate representation of it.

The convergence of the identification procedure is proved by the direct application of the ergodic theorem. This chapter then depends only on the results of Chapter 3 and Section 4.1, while the more advanced probabilistic arguments used in Chapter 4 are omitted. (Note that since we consider here a very specific class of observation sequences, we are, in fact, able to treat a more interesting class of problems than that considered in Section 4.2 as the true parameter is not assumed to belong to the parameter set.)

5.1 Linear Neural Circuits

Consider a neural circuit of the mathematical form

$$x_{n+1} = F_* x_n + G_* w_n$$
$$z_n = H_* x_n + \nu_n \tag{5.1a}$$

(where, as noted in Chapter 2, * marks a true parameter) initialized at $n = n_0$ with

$$E x_{n_0} = 0; \quad E x_{n_0} x_{n_0}^T = \Psi_{*,n_0}$$

where w_n and ν_n are uncorrelated and mutually uncorrelated Gaussian sequences with

$$E w_n = E \nu_n = 0;$$
$$E\{w_n\, w_n^T\} = Q_*; \quad E\{\nu_n\, \nu_n^T\} = R_* \tag{5.1b}$$

The above neural circuit is represented by the divided set

$$M_1 \equiv \{(F_j, G_j, H_j, Q_j, R_j); j \in K \equiv (0, 1, \ldots, p)\} \tag{5.2}$$

which represents a group of Brain Proximity Measures.

Let

$$K' \equiv (* \cup K)$$

(As in Chapter 4, the restriction to a finite set is done for the analysis of convergence and consistency. In Section 5.4, we consider other modeling problems, and there the model set is allowed to be infinite. Also, note that the results of the present chapter can easily be extended to the case where the system (5.1a) is driven by an additional deterministic input sequence.)

Let

$$\hat{z}_{j,n} \equiv E_j^{\cup_{n-1}} z_n; \quad j \in K'$$

denote the one-step least-squares prediction of z_n, given the past observations z^{n-1}, assuming that the jth model is the true one.

For each $i, j \in K'$, let

$$\Sigma_{j,n} = \Sigma_j(n, n_0) \equiv E_j\{(z_n - \hat{z}_{j,n})^T(z_n - \hat{z}_{j,n})\} \tag{5.3}$$

and

$$\Gamma^i_{j,n} \equiv \Gamma^i_j(n, n_0) \equiv E_i\{(z_n - \hat{z}_{j,n})^T(z_n - \hat{z}_{s,n})\} \tag{5.4}$$

Denote further

$$\Sigma_j \equiv \lim_{n_0 \to -\infty} \Sigma_j(n, n_0) \tag{5.5}$$

provided that the limit exists.

We shall use the following condition:

(c5.1) For each $j \in K', \Sigma_j$ exists and has a finite positive definite value.

A sufficient condition for (c5.1) is that each model corresponding to $j \in K'$ is detectable and controllable. For each $j \in K', \Sigma_j$ is obtained by running to steady state a Riccati equation corresponding to $(F_j, G_j, H_j, Q_j, R_j)$.

Also, denote

$$\Gamma^t_s \equiv \lim_{n_0 \to -\infty} \Gamma^i_j(n, n_0) \tag{5.6}$$

provided that the limit exists. Γ^i_j is obtained by the following procedure. First, assuming that $n_0 = -\infty$, take $\Sigma_{j,n} = \Sigma_j$ for each $j \in K'$ and for all $n \geq n_1$, where n_1 is any fixed integer. Then the dynamic equation generating simultaneously, according to the measure P_i, the state x_n and its one-step prediction corresponding to $\hat{x}_{j,n}$, is

$$\begin{bmatrix} x_{i,n+1} \\ \hat{x}_{j,n+1} \end{bmatrix} = \begin{bmatrix} F_i & 0 \\ F_j K_j H_i & F_j(I - K_j H_j) \end{bmatrix} \begin{bmatrix} x_{i,n} \\ \hat{x}_{j,n} \end{bmatrix} + \begin{bmatrix} G_i & 0 \\ 0 & F_j K_j \end{bmatrix} \begin{bmatrix} w_n \\ v_n \end{bmatrix}$$

where

$$K_j = \Sigma_j H_j^T (H_j \Sigma_j H_j^T + R_j)^{-1}$$

Let

$$F^i_j \equiv \begin{bmatrix} F_i & 0 \\ F_j K_j H_j & F_j(I - K_j H_j) \end{bmatrix}; \quad G^i_j \equiv \begin{bmatrix} G_i & 0 \\ 0 & F_j K_j \end{bmatrix}$$

and

$$Q^i \equiv \begin{bmatrix} Q_i & 0 \\ 0 & R_i \end{bmatrix}; \quad H_j^i \equiv [H_i - H_j]$$

Then the matrix

$$\Psi_{j,n}^i \equiv E\left\{ \begin{bmatrix} x_{i,n+1} \\ \hat{x}_{j,n+1} \end{bmatrix} [x_{i,n+1} \quad \hat{x}_{j,n+1}] \right\}$$

is generated by the Lyapunov equation

$$\Psi_{j,n+1}^i = F_j^i \Psi_{j,n}^i F_j^{i^T} + G_j^i Q^i G_j^i \tag{5.7}$$

initiated at n_1 by any initial value. We can write

$$\Psi_{j.n}^i = \Psi_j^i(n, n_1)$$

Then let

$$\Psi_j^t = \lim_{n_1 \to -\infty} \Psi_j^i(n, n_1) \tag{5.8}$$

Finally,

$$\Gamma_j^i = H_j^i \Psi_j^i H_j^{i^T} + R_i \tag{5.9}$$

It is well known that the limit in Eq. (5.8) of Eq. (5.7) exists and is finite if the matrix F_j^i has all its eigenvalues inside the unit circle. This is the case if for each $j \in K', F_j$ has all its eigenvalues inside the unit circle and (F_j, H_j) is observable. Note, however, that these conditions are only sufficient, not necessary, for Γ_j^i to be finite, since Eq. (5.9) may be finite even if Ψ_j^i obtained as the limit value of Eq. (5.7) is not finite.

Theorem 5.1. *For each $j \in K'$, let the corresponding model be stable and observable and let $n_0 = -\infty$. Then the residuals sequences $(z_n - \hat{z}_{j,n}); j \in K'; n \geq 0$ are ergodic according to the true probability measure.*

Proof. We have, by Eq. (5.5), $\sum_{j,n} = \sum_j$ for all $n \geq 0$. Since both (z_n) and $(\hat{z}_{j,n})$ are linear operators on a zero mean Gaussian sequence (x_n), they are zero-mean Gaussian, and so is the sequence

$(z_n - \hat{z}_{n,j})$ for each $j \in K'$. Hence $(z_n - \hat{z}_{j,n})$ is a zero-mean stationary Gaussian sequence for each $j \in K'$. By Proposition 2.1, we have that $(z_n - \hat{z}_{j,n})$ is ergodic if and only if

$$\lim_{n\to\infty} \frac{1}{n+1} \sum_{k=0}^{n} |R(K)|^2 = 0 \tag{5.10}$$

where $|R(k)|$ denotes the determinant of the matrix

$$R(k) = E\{(z_n - \hat{z}_{j,n})(z_{n+k} - \hat{z}_{j,n+k})^T\} = \begin{cases} H_j^i \Psi_j^i H_j^{i^T} + R_i & k = 0 \\ H_j^i \Psi_j^i H_j^{i^T} + F_j^{i^k} & k > 0 \end{cases} \tag{5.11}$$

We have, for any $k > 0$,

$$|R(k)| = |H_j^i H_j^{i^T}| |\Psi_j^i|^k |F_j^i|^k$$

Since all eigenvalues of F_j^i are inside the unit circle, then

$$|F_j^i| < 1$$

Hence,

$$\begin{aligned} \lim_{n\to\infty} \sum_{k=1}^{n} |R(k)|^2 &= |H_j^i H_j^{i^T}|^2 |\Psi_j^i|^2 \lim_{n\to\infty} \sum_{k=1}^{n} |F_j^i|^{2k} \\ &= \frac{|H_j^i H_j^{i^T}|^2 |\Psi_j^i|^2}{1 - |F_j^i|^2} < \infty \end{aligned} \tag{5.12}$$

yielding

$$\lim_{n\to\infty} \frac{1}{n+1} \sum_{k=0}^{n} |R(k)|^2 = 0$$

The assertion follows. □

Note that the stability and observability of the models, assumed in Theorem 5.1, are only sufficient, not necessary. In fact, we have proved ergodicity of the state residuals $(x_n - \hat{x}_{j,n})$, which is not a necessary condition, to show the ergodicity of $(z_n - z_{j,n})$. In the sequel, we shall directly use the following condition:

(c5.2) For each $j \in K'$, the residuals sequence $(z_n - \hat{z}_{j,n})$ is ergodic.

5.2 Information, Convergence and Consistency

Consider the system in Eq. (5.1) and the model set in Eq. (5.2). Let condition (c5.1) hold. Then the conditional probability density of z_n, given the past observations Z^{n-1} corresponding to each model, is

$$f_j(z_n|Z^{n-1}) = [(2\pi)^\ell|\Sigma_j|]^{-1/2}\exp\left\{-\frac{1}{2}(z_n-\hat{z}_{j,n})^T\Sigma_j^{-1}(z_n-\hat{z}_{j,n})\right\}; \quad j \in K' \tag{5.13}$$

where ℓ is the dimension of z_n.

In Chapter 3, we defined, for each pair $j, k \in K'$,

$$\overline{I}_n(k;j) \equiv E_* \log\frac{f_k(z_n|Z^{n-1})}{f_j(z_n|Z^{n-1})}$$

and

$$d_n(k;j) \equiv |\overline{I}_n(k;j)|$$

We have, for each $j \in K'$,

$$E_* \log f_j(z_n|Z^{n-1}) = -\frac{\ell}{2}\log 2\pi - \frac{1}{2}\log|\Sigma_j| - \frac{1}{2}\mathrm{tr}\Sigma_j^{-1}\Gamma_j^* \tag{5.14}$$

and for each pair $j, k \in K'$,

$$\overline{I}_n(k;j) = \overline{I}(k;j) = \frac{1}{2}\log|\Sigma_j| + \frac{1}{2}\mathrm{tr}\Sigma_j^{-1}\Gamma_j^* - \frac{1}{2}\log|\Sigma_k| - \frac{1}{2}\mathrm{tr}\Sigma_k^{-1}\Gamma_k^* \tag{5.15}$$

Let

$$L_j^i \equiv \log|\Sigma_j| + \mathrm{tr}\Sigma_j^{-1}\Gamma_j^i \quad i, j \in K' \tag{5.16}$$

Then we have

$$\overline{I}_n(k;j) = \frac{1}{2}(L_j^* - L_k^*) \quad j, k \in K \tag{5.17}$$

Also, by Theorem 3.1,

$$\overline{I}_n(*;j) \geq 0 \quad \text{for each } j \in K$$

Hence,

$$d(*;j) = d_n(*;j) = \overline{I}_n(*;j) \quad \text{for each } j \in K$$

Thus, for any $j, k \in K$

$$d(*;j) - d(*;k) = \overline{I}_n(*;j) - I_n(*;k) = I_n(k;j) = \frac{1}{2}(L_j^* - L_k^*) \tag{5.18}$$

Hence,

$$d(*;j) \geq d(*;k)$$

if and only if

$$L_j^* \geq L_k^*$$

Lemma 5.1. *Let* (z_n) *be generated by Eq.* (5.1) *and let condition* (c5.1) *hold. Then, under condition* (c5.2), *for any* $j, k \in K$,

$$\lim_{n\to\infty} h_k^j(Z^n) = 0 \quad \text{a.s.} \tag{5.19}$$

if

$$L_k^* < L_j^* \tag{5.20}$$

and only if

$$L_k^* \leq L_j^* \tag{5.21}$$

Proof. We have

$$\log h_k^j(z^n) = \sum_{m=1}^{n} \log h_k^j(z_m|Z^{m-1}) \tag{5.22}$$

yielding

$$\log h_k^j(z_n|Z^n) = \frac{1}{2}\log\frac{|\Sigma_k|}{\Sigma_j} - \frac{1}{2}(z_n - \hat{z}_{j,n})^T\Sigma_j^{-1}(z_n - \hat{z}_{j,n})$$
$$+ (z_n - \hat{z}_{k,n})^T\Sigma_k^{-1}(z_n - \hat{z}_{k,n}) \tag{5.23}$$

Since for each $j \in K$, the residuals $(z_n - \hat{z}_{j,n})$ are ergodic, it follows from the ergodic theorem (Theorem 2.5) that

$$\lim_{n\to\infty}\frac{1}{n}\sum_{m=1}^{n}\log h_k^j(z_m|Z^{m-1}) = E_*\log h_k^j(z_m|Z^{m-1}) \quad \text{a.e.}$$
$$= \overline{I}_m(j;k) = \frac{1}{2}(L_k^* - L_j^*) \tag{5.24}$$

Now, if

$$L_k^* < L_j^* \tag{5.25}$$

then obviously

$$\lim_{n\to\infty} \sum_{m=1}^{n} \log h_k^j(z_m|Z^{m-1}) = \lim_{n\to\infty} \log h_k^j(Z^n) = -\infty \quad \text{a.e.}$$

yielding

$$\lim_{n\to\infty} h_k^j(Z^n) = 0$$

To prove that Eq. (5.19) implies (5.21), suppose that Eq. (5.19) holds, but Eq. (5.21) does not, then

$$L_k > L_j \tag{5.26}$$

and by Eq. (5.24),

$$\lim_{n\to\infty} \sum_{m=1}^{n} \log h_k^j(z_m|Z^{m-1}) = \lim_{n\to\infty} \log h_k^j(Z^n) = \infty$$

implying

$$\lim_{n\to\infty} h_k^j(Z^n) = \infty \tag{5.27}$$

which contradicts Eq. (5.19). Thus, Eq. (5.19) implies Eq. (5.21). □

Consider the following condition:

(c5.3) There exists some parameter $k \in K$ such that

$$L_k^* < L_j^* \quad \text{for all } j \in K; \quad j \neq k \tag{5.28}$$

Theorem 5.2. *Consider the system in Eq.* (5.1) *and the model set* M_1, *and let* (c5.1) *hold. Under conditions* (c5.2) *and* (c5.3), *the ML, the MAP and the LS identification methods will converge a.e. and in the mean square to the model set* $(F_k, G_k, H_k, Q_k, R_k)$.

Proof. By Lemma 5.1, condition (c4.1) is satisfied for the parameter k. The assertion then follows from Theorems 4.1–4.4. □

Note that by Eq. (5.18), the identified model is the closest to the true model in the metric d.

Corollary 5.1. *The convergence specified in Theorem* 5.2 *will be to a model in the model set* M_1 *such that*

$$|L_k^* - L_*^*| = \min\{|L_j^* - L_*^*|; j \in K\} \tag{5.29}$$

Proof. By Theorem 3.1, we have

$$\bar{I}(*; j) \geq 0 \quad \text{for each } j \in K$$

Hence,

$$L_j^* - L_*^* > 0$$

if

$$L_j^* \neq L_*^*$$

So, the assertion follows from Lemma 5.1 and Theorem 5.2. □

The identification methods will then converge to a parameter in K, closest to the true parameter in the scalar L, which, in turn, implies closeness of the corresponding models in terms of their output statistics.

Corollary 5.2. *Suppose that the true system belongs to the set* M_1 *i.e., let* $(F, G, H, Q, R) = (F_r, G_r, H_r, Q_r, R_r)$ *for some* $r \in K$. *Let conditions* (c5.1) *and* (c5.2) *hold and suppose that for each* $j \in K; j \neq r$, *we have*

$$L_j^* = L_j^r \neq L_r^r = L_*^* \tag{5.30}$$

Then the identification procedure will converge to the true model in the sense specified in Theorem 5.2.

Proof. The result follows immediately from Corollary 5.1. □

To compute the scalars $L_j^*, j \in K$, one must compute the matrices Σ_j and Γ_j^*. While the matrix Σ_j can be computed by running a Riccati equation corresponding to the jth model to steady state, the matrix Γ_j^* cannot be computed unless the true measure or, equivalently, the true system, is known. If $r \in K$ is the true parameter, then

$$\Gamma_r^* = \Gamma_r^r = \Sigma_r$$

and, consequently,

$$L_r^* = L_r^r = \log|\Sigma_r| + \ell \tag{5.31}$$

In the identification problem, the true parameter is unknown. If the true parameter can be assumed to belong to the parameter set, then Eq. (5.30) will have to be checked for all pairs of parameters in the set, i.e.,

(c5.4) For all pairs $i, j \in K, i \neq j$,

$$L_j^i \neq L_i^i = \log|\Sigma_i| + \ell \tag{5.32}$$

Theorem 5.3. *Let the system in Eq.* (5.1) *belong to the set* M_1, *and let conditions* (c5.1) *and* (c5.2) *hold. Then, under condition* (c5.4), *the true model is identifiable a.e. by the ML and MAP estimates, identifiable a.e. and in the mean square by the LS estimate.*

Proof. Under condition (c5.4), we have Eq. (5.30). The assertion then follows directly from Corollary 5.2. □

5.3 L_1 Convergence

We have shown in Section 4.4 the L_1 convergence of the likelihood ratios and the *a posteriori* probability ratios under condition (c4.1). Furthermore, it was shown that under condition (c4.5), bounds on the L_1 convergence rates can be established, thus providing a measure of performance of the ML and MAP estimation methods. We now show L_1 convergence and derive L_1 convergence bounds for the identification of stationary linear systems addressed in the present chapter.

Consider the system in Eq. (5.1) and the model set M_1 and let condition (c5.1) hold. We have shown (Eq. (5.17)) that under condition (c5.2),

$$\overline{I}_n(k;j) = \overline{I}(k;j) = \frac{1}{2}(L_j^* - L_k^*) \quad \text{for all } n$$

for each pair $k, j \in K$, where, $L_j^*; j \in K$ are constants.

Theorem 5.4. *Consider the system in Eq.* (5.1) *and the model set* M_1 *in Eq.* (5.2). *Under conditions* (c5.1) *and* (c5.3), *for each* $j \in K$; $j \neq k$, *the sequences* $(h_j^k(Z^n))$ *and* $\left(\frac{f^b(k|Z^n)}{f^b(j|Z^n)}\right)$ *converge in* L_1 *to infinity. Furthermore, the sequences converge at rates no slower than exponential.*

Proof. By Lemma 5.1, condition (c5.3) implies condition (c4.1). The L_1 convergence of both sequences follows from Theorem 4.10.

Now, let

$$\alpha_j \equiv \frac{1}{2}(L_j^* - L_k^*) \quad \text{for each } j \in k : j \neq k$$

then following the proof of Theorem 4.11, we get by Eqs. (4.11) and (4.12), for each $j \in k : j \neq k$,

$$Eh_j^k(Z^n) \geq e^{(n+1)\alpha_j} \tag{5.33}$$

and

$$E\frac{f^b(k|Z^n)}{f^b(j|Z^n)} \geq \frac{f^b(k)}{f^b(j)}e^{(n+1)\alpha_j} \tag{5.34}$$

□

The rates $\alpha_j = \overline{I}(k;j); j \in K; j \neq k$ can only be computed, as discussed in Section 5.2, if the true model is known. If the true model is only known to belong to the set M_1, then the rates can be bounded as follows. We have seen that if $k \in K$ is the true parameter, then (see Eq. (5.31))

$$L_k^* = \log|\Sigma_k| + \ell$$

Now, since

$$\alpha_j = \frac{1}{2}(L_j^* - L_k^*) \quad \text{for each } j \in K; \quad j \neq k$$

where k is the true parameter, we have

$$\alpha_j = \frac{1}{2}(L_j^* - \log|\Sigma_k| - \ell)$$

Then

$$\alpha_j \geq \alpha \equiv \min\left\{\min\left\{\frac{1}{2}[L_j^k - \log|\Sigma_k| - \ell]; \quad L_j^k - \log|\Sigma_k| - \ell = 0 \quad \text{for all } j \in K; \quad j \neq k;\right\} k \in K\right\} \tag{5.35}$$

Equation (5.35) reads as follows: For each $k \in K$, suppose that k is the true parameter. If

$$L_j^k - \log|\Sigma_k| - \ell \geq 0 \quad \text{for all } j \in K; \quad j \neq k \tag{5.36}$$

then take the min over j of Eq. (5.36). Continue the procedure over all $k \in K$, discarding such k for which Eq. (5.36) does not hold (since then k cannot be the true parameter, for which Eq. (5.36) always holds). Then take the least of all the minimum values of Eq. (5.26) found above. Note that this procedure does not identify the true parameter, but, rather, finds a lower bound for α_j over $j \in K$.

The above discussion is summarized in the following theorem.

Theorem 5.5. *Consider the system in Theorem* 5.1 *and suppose that its true model belongs to the set* M_1 *given by Eq.* (5.2). *Then, under condition* (c5.3), *we have*

$$E_j^r(Z^n) \geq e^{(n+1)\alpha} \tag{5.37}$$

$$E\frac{f^b(r|Z^n)}{f^b(j|Z^n)} \geq \frac{f^b(r)}{f^b(j)}e^{(n+1)\alpha} \tag{5.38}$$

for each $j \in K; j \neq r$, *where* r *is the true parameter and* α *is given by Eq.* (5.35).

As discussed in Section 4.4, the bounds, Eqs. (5.37) and (5.38), provide performance measures for the ML and MAP estimation methods. We have shown that bounds on the L_1 convergence rates of the indicated ratios can actually be computed for stationary Gaussian linear systems.

5.4 Model Selection

In practice, when a mathematical model of a dynamical system is required for purposes of estimation and control, one often knows, to a certain approximation, what the model should be. However, because of implementation constraints, one has to select a different model. Such is the case when the actual system is of high order, but the available computation and storage capabilities are such that only a low-order model can be handled. Another modeling problem arises when the actual system's model is known to take its values, which may be time-varying, from a given set, but only a single model can be considered. These are not identification problems in the strict sense. Nevertheless, the analysis in the previous sections suggests a natural extension of the results to the model selection problems introduced above. It should be emphasized that unlike the investigation of convergence and consistency of parameter estimates, the following discussion applies to infinite and even non-compact parameter sets.

5.4.1 *The selection of a reduced order model*

Suppose that the true system, or an approximate model of it, is known, but their dimensions are too high for implementation of estimation and control procedures. A model of a lower dimension is then desired. Let the true system, or an approximate model of it, be given by (5.1) and let

$$M \equiv (F_s, G_s, H_s, Q_s, R_s); \quad s \in S \tag{5.39}$$

be a model set of dimension lower than that of Eq. (5.1). The system coefficients in M depend on a parameter vector s belonging to a parameter set S. It is desired to find the model in the set M which is closest to the true system $(F_*, G_*, H_*, Q_*, R_*)$ in some meaningful distance sense.

For each $s \in S$, let

$$L_s^* \equiv \log|\Sigma_s| + \mathrm{tr}\Sigma_s^{-1}\Sigma_s^* \tag{5.40}$$

where

$$\Gamma_s^* \equiv E_*\{(z_n - \hat{z}_{s,n})(z_n - \hat{z}_{s,n})^T\}$$

$\hat{z}_{s,n}$ is the one-step least-square prediction of z_n given the past observations Z^{n-1}, assuming that s is the true parameter value and Σ_s is the corresponding prediction error covariance matrix. Σ_s is obtained by running a Riccati equation corresponding to the model $(F_s, G_s, H_s, Q_s, R_s)$ to steady state. Let $s^\circ \in S$ be a parameter which satisfies the following criterion:

$$L^*_{s_\circ} < \{L^*_s; s \in S, s \neq s^\circ\}$$

Then, following the reasoning of Section 5.2, the model $(F_s, G_s, H_s, Q_s, R_s)$ satisfies the following equivalent criteria:

(1) the model that is closer to the true model than any other model in the sense $d(*; s^\circ) \leq \{d(*; s); s \in S\}$,
(2) the model that would be favored over any other model in M by the incoming information,
(3) the model that would be identified as the true model among any finite set of models from the set M by the maximum likelihood and Bayesian estimation techniques.

The model selection problem reduces then to the minimization problem

$$\min_s \{L^*_s; s \in S\} \tag{5.41}$$

5.4.2 *The selection of a representative model*

Suppose that the model of a linear system whose parameters may be time-varying is known to take its values from a set

$$M \equiv (F_s, G_s, H_s, Q_s, R_s); \quad s \in S$$

Two different cases may be considered:

(1) The model takes a certain constant value in the set M, and there is no prior knowledge, even in a probabilistic sense, on what value it might be.
(2) During the system's operation, its mathematical model varies over the model set M. However, the model's variation is unknown or is impossible to handle because of storage or computation limitations.

In either case, it is desired to select a single model from the set M to represent the system throughout its operation. One criterion for the selection of such a model is that the maximum possible distance d between the representative model and the true model (whatever it might be) will be minimal.

The procedure for selecting the representative model from M will then be as follows. First, for each parameter, $s \in S$, find the parameter t whose distance from s is maximal. Then, find the parameter s for which the maximal distance found in the first step is minimal.

The distance between a parameter s and the parameter t of the set S is maximized over t by maximizing with respect to t the following term:

$$L_s^t = \log|\Sigma_s| + \mathrm{tr}\Sigma_s^{-1}\Gamma_s^t \tag{5.42}$$

where, as before,

$$\Sigma_s \equiv E_s\{(z_n - \hat{z}_{s,n})(z_n - \hat{z}_{s,n})^T\}$$

is obtained by

$$\Gamma_s^t \equiv E_t\{(z_n - \hat{z}_{s,n})(z_n - \hat{z}_{s,n})^T\}$$

which is obtained by running a Lyapunov equation to steady state, as shown in the previous section.

The representative model is then found by solving the minmax problem

$$\min_s \max_t \{L_s^t; s, t \in \zeta\} \tag{5.43}$$

5.5 Concluding Remarks

The procedures described in this chapter find, in general, a model in the model set, whose output statistics are best matched with those of the true system. However, for the modeling problem considered, the role of the output can be played by any linear function of the state variables. If, for instance, it is desired to emphasize certain variables that affect the system's performance more than the others or that can be measured better than the others, then these variables can be selected as outputs for the model selection procedures.

The problem of selecting a single model from a model set considered in this chapter can be generalized to a problem of selecting any number of models from the set, so that the model set is approximately represented by a finite set of models. An identification procedure can then be employed "online" to find the model in the finite set which is closest to the true system. The selection of a finite model set would require, as a first step, the division of the infinite parameter set into a finite number of subsets. The way in which the parameter space should be divided would depend on considerations of the physical problem involved, but it seems obvious that the division would employ the metric topology of the parameter space (just as interval lengths are used in R^n to divide a rectangle into equal parts).

Chapter 6

Non-Stationary Linear Systems

The assumption of stationary systems made in the previous chapter is now removed, as we consider the more general case of non-stationary, time-varying linear systems. Yet, as in the case of stationary linear systems, the analysis is centered on finite sets of cortical neural circuits. We first derive expressions for the information in the observations, discriminating one model in the model set against the others. The information conditions for the consistency of the estimates are interpreted in terms of the second-order statistics associated with the different neural circuits and computed by solving the corresponding Riccati equations. The consistency result for time-varying systems is not, however, as explicit as in the stationary case. The L_1 convergence of the likelihood and the *a posterior* probability ratios are investigated. The separate contributions of the stochastic and the deterministic parts of the input to the information and, consequently, to the L_1 convergence rates, are shown.

6.1 Models

Consider the system

$$\begin{aligned} x_{n+1} &= F_{*,n} x_n + G_{*,n} w_n \\ z_n &= H_{*,n} x_n + v_n \end{aligned} \tag{6.1a}$$

initialized at $n = n_0$ with

$$E_* x_{n_0} = 0; \quad E_* x_{n_0} x_{n_0}^T = \Psi_*$$

where w_n and ν_n are uncorrelated and mutually uncorrelated Gaussian sequences with

$$Ew_n = E\nu_n = 0$$
$$E\{w_n w_n^T\} = Q_{*,n}; \quad E\{v_n v_n^T\} = R_{*,n} \tag{6.1b}$$

Consider a finite set of cortical neural circuits

$$M_2 \equiv \{(F_{j,n}, G_{j,n}, H_{j,n}, \Psi_j, Q_{j,n}, R_{j,n}); j \in K = (0, 1, \ldots, p)\} \tag{6.2}$$

which represents a group of Brain Proximity Measures.

Let z_n be an ℓ-dimensional observation sequence. The conditional probability density of z_n given the past observations Z^{n-1} and corresponding to each model is given by

$$f_j(z_n|Z^{n-1})$$
$$= \left[(2\pi)^\ell \left|\sum_{j,n}\right|\right]^{-1/2} \exp\left\{-\frac{1}{2}(z_n - \hat{z}_{j,n})^T \sum_{j,n}^{-1}(z_n - \hat{z}_{j,n})\right\};$$
$$j \in k \tag{6.3}$$

where, as before, $\hat{z}_{j,n}$ is the one-step prediction of z_n given the past observations Z^{n-1}, assuming that the jth model is the true one (i.e., assuming that the observations are generated by the jth model) and $\sum_{j,n}$ is the corresponding error covariance matrix.

6.2 Information, Convergence and Consistency

The information in a single observation z_n, favoring the kth model against the jth model, is derived next:

$$I_n(k;j) = E_*^{U_{n-1}} \log \frac{f_k(z_n|Z^{n-1})}{f_j(z_n|Z^{n-1})}$$
$$= E_*^{U_{n-1}} \left\{-\frac{1}{2}\log \sum_{k,n} - \frac{1}{2}(z_n - \hat{z}_{k,n})^T \sum_{k,n}^{-1}(z_n - \hat{z}_{k,n})\right\}$$
$$+ \frac{1}{2}\log \sum_{j,n} + \frac{1}{2}(z_n - \hat{z}_{k,n})^T \sum_{k,n}^{-1}(z_n - \hat{z}_{k,n})$$

$$= \frac{1}{2}\log\frac{|\sum_{j,n}|}{|\sum_{k,n}|} - \frac{1}{2}\mathrm{tr}\sum_{k,n}^{-1} E_*^{U_{n-1}}\{z_n z_n^T\} + \frac{1}{2}\hat{z}_{k,n}^T \sum_{k,n}^{-1} \hat{z}_{*,n}$$

$$+ \frac{1}{2}\hat{z}_{*,n}^T \sum_{k,n}^{-1} \hat{z}_{k,n} - \frac{1}{2}\hat{z}_{*,n}^T \sum_{k,n}^{-1} \hat{z}_{*,n} + \frac{1}{2}\mathrm{tr}\sum_{j,n}^{-1} E_*^{U_{n-1}}\{z_n z_n^T\}$$

$$- \frac{1}{2}\hat{z}_{j,n}^T \sum_{j,n}^{-1} \hat{z}_{*,n} - \frac{1}{2}\hat{z}_{*,n}^T \sum_{j,n}^{-1} \hat{z}_{j,n} + \frac{1}{2}\hat{z}_{*,n}^T \sum_{j,n}^{-1} \hat{z}_{*,n} \quad \text{a.e.}$$

But

$$E_*^{U_{n-1}}\{z_n z_n^T\} = \sum_{*,n} + \hat{z}_{*,n}\hat{z}_{*,n}^T \quad \text{a.e.}$$

Hence,

$$I_n(k;j) = \frac{1}{2}\log\frac{|\sum_{j,n}|}{|\sum_{k,n}|} + \frac{1}{2}\mathrm{tr}\sum_{*,n}\left(\sum_{j,n}^{-1} - \sum_{k,n}^{-1}\right) - \frac{1}{2}(\hat{z}_{*,n} - \hat{z}_{k,n})^T$$

$$\times \sum_{k,n}^{-1}(\hat{z}_{*,n} - \hat{z}_{k,n}) + \frac{1}{2}(\hat{z}_{*,n} - \hat{z}_{j,n})^T \sum_{j,n}^{-1}(\hat{z}_{*,n} - \hat{z}_{j,n}) \quad \text{a.e.} \tag{6.4}$$

Let

$$I_n^{(1)}(k;j) \equiv \frac{1}{2}\log\frac{|\sum_{j,n}|}{|\sum_{k,n}|} + \frac{1}{2}\mathrm{tr}\sum_{*,n}\left(\sum_{j,n}^{-1} - \sum_{k,n}^{-1}\right) \tag{6.5}$$

and

$$I_n^{(2)}(k;j) \equiv -\frac{1}{2}(\hat{z}_{*,n} - \hat{z}_{k,n})^T \sum_{k,n}^{-1}(\hat{z}_{*,n} - \hat{z}_{k,n})$$

$$+ \frac{1}{2}(\hat{z}_{*,n} - \hat{z}_{j,n})^T \sum_{j,n}^{-1}(\hat{z}_{*,n} - \hat{z}_{j,n}) \tag{6.6}$$

Hence,

$$I_n(k;j) = I_n^{(1)}(k;j) + I_n^{(2)}(k;j) \tag{6.7}$$

Suppose that condition (c4.2) is satisfied. Then, by Lemma 4.2 and by Theorems 4.1–4.4, conditions (c4.3) or (c4.4) are sufficient for convergence to the kth circuit in M_1. However, it is not difficult to see that the verification of conditions (c4.3) or (c4.4) is not possible for the general case considered here under any conditions imposed on the deterministic part $I_n^{(1)}(k;j)$ due to the random part $I_n^{(2)}(k;j)$. In Section 6.3, we shall show that $I_n^{(2)}(k;j)$ can be further separated into deterministic and stochastic parts. We now show that under the assumption that the true model belongs to the given set M_2, the information expression for the time-varying system under consideration is simplified, and consequently, some explicit conditions for identification can be obtained.

Suppose that some $k \in k$ is the true parameter, then by Eq. (6.4),

$$I_n(k;j) = I'_n(k;j) + I''_n(k;j) \tag{6.8}$$

where

$$I'_n(k;j) \equiv \frac{1}{2}\log\frac{|\sum_{j,n}|}{|\sum_{k,n}|} + \frac{1}{2}\mathrm{tr}\left(\sum_{k,n}\sum_{j,n}^{-1} - I\right) \tag{6.9}$$

and

$$I''_n(k;j) \equiv \frac{1}{2}(\hat{z}_{k,n} - \hat{z}_{j,n})^T \sum_{j,n}^{-1}(\hat{z}_{k,n} - \hat{z}_{j,n}) \tag{6.10}$$

Consider the following condition:

(c6.1) For some $k \in K$ and for each $j \in K$; $j \neq k$, there exists some scalar $\alpha_j > 0$ and a subsequence (n^j) of (n) such that

$$\left\|\sum_{k,n^j} - \sum_{j,n^j}\right\| \geq \alpha_j \quad \text{for all } n^j \tag{6.11}$$

where

$$\|A\| \equiv |\det A|$$

Lemma 6.1. *Let some $k \in K$ be the true parameter, i.e., let*

$$\begin{aligned}(F_{*,n}, G_{*,n}, H_{*,n}, \Psi_{*,n}, Q_{*,n}, R_{*,n}) \\ = (F_{k,n}, G_{k,n}, H_{k,n}, \Psi_{k,n}, Q_{k,n}, R_{k,n})\end{aligned}$$

Then condition (c6.1) *implies condition* (c4.4) *for k.*

Proof. Clearly,

$$I''_n(k;j) \geq 0 \quad \text{for all } n \text{ for each } j \in K$$

Thus,

$$I_n(k;j) \geq I'_n \quad \text{for all } n \text{ for each } j \in K$$

It will suffice then to show that condition (c6.1) implies the existence of a subsequence (n_r^j) of (n^j) and some $\varepsilon_j > 0$ such that

$$I'_{n_r^j}(k;j) \geq \varepsilon_j \quad \text{for all } n_r^j \tag{6.12}$$

Consider the following equation:

$$\left| \sum_{k,n} - \lambda_n \sum_{j,n} \right| = 0 \tag{6.13}$$

For positive definite $\sum_{k,n}$ and $\sum_{j,n}$, there exists a nonsingular matrix A_n such that (Anderson, 1958, p. 341)

$$A_n^T \sum_{k,n} A_n = \Lambda_n \tag{6.14}$$

and

$$A_n^T \sum_{j,n} A_n = I \tag{6.15}$$

where Λ_n is a diagonal matrix whose elements are $\lambda_{n,i}, i = 1, \ldots, \ell$ the roots of Eq. (6.13). In addition, we have $\lambda_{n,i} \geq 1$ for all $i = 1, \ldots, \ell$, and $n \geq 0$.

It is easy to verify that $I'(k;j)$ remains invariant under the transformation in Eqs. (6.14) and (6.15). Hence,

$$\begin{aligned} I'_n(k;j) &= -\frac{1}{2} \log |\Lambda_n| + \frac{1}{2} \operatorname{tr}(\Lambda_n - I) \\ &= \frac{1}{2} \sum_{i=1}^{\ell} [\lambda_{n,i} - \log \lambda_{n,i} - 1] \end{aligned} \tag{6.16}$$

Suppose that for some subsequence (n^j) of (n),

$$\left\| \sum_{k,n_j} - \sum_{j,n_j} \right\| \geq \alpha_j > 0 \quad \text{for all } n^j \tag{6.17}$$

Then there exists some $\varsigma_j > 0$ and a subsequence (n_r^j) of (n^j) such that

$$\|\lambda_{n_r^j,i} - 1\| \geq \varsigma_j \quad \text{for all } n_r^j \quad \text{for each } i = 1, \ldots, \ell \tag{6.18}$$

since if such ς_j and n_r^j do not exist, then

$$\lambda_{n^j}, i_{n^j} \to 1 \quad \text{as} \quad n \to \infty \tag{6.19}$$

where

$$\lambda_{n^j}, i_{n^j} \equiv \min\{\lambda_{n^j}; i = 1, \ldots, \ell\}$$

and then

$$\left| \sum_{k,n_r^j} - \sum_{j,n_r^j} \right| \to \left| \sum_{k,n_r^j} - \lambda_{n_r^j}, i_{n_r^j} \sum_{j,n_r^j} \right| = 0$$

as $n_r^j \to \infty$, contradicting Eq. (6.17). Hence, Eq. (6.17) implies Eq. (6.18).

Now, consider Eq. (6.16). Since

$$a - \log a - 1 \geq 0 \tag{6.20}$$

with equality if and only if $a = 1$, and since the function on the left-hand side of Eq. (6.20) is convex in a, it follows that given $\zeta > 0$, there exists some $\alpha > 0$ such that

$$a - \log a - 1 > \alpha$$

whenever $|a - 1| \geq \varsigma$.

Thus, finally, Eq. (6.18) implies that there exists some $\varepsilon_j > 0$ such that

$$I'_{n_r^j}(k; j) \geq \varepsilon_j \quad \text{for all } n_r^j \tag{6.21}$$

The assertion follows. □

We have shown in Chapter 4 that consistency of the parameter estimate (or, equivalently, identifiability of the dynamical system) follows from conditions (c4.2) and (c4.4). Condition (c4.4) (or, more generally, (c4.3)) seems to be, for obvious reasons, the "crucial" condition for the strong consistency of the estimates. We show in the following that condition (c4.2) holds for the case of time-invariant stationary linear systems. It seems, however, that condition (c4.2) would hold for very general classes of observation sequences. For the general case of time-varying systems, we condition the consistency result on condition (c4.2), which has to be checked for each case under consideration. It seems, in particular, that condition (c4.2) would not be difficult to verify for the class of periodically varying linear systems and for systems driven by bounded deterministic inputs.

Theorem 6.1. *Suppose that the system* (6.1) *belongs to the set* M_2 *specified by* (6.2). *Furthermore, suppose that condition* (c4.2) *holds. Then the system is identifiable a.e. by the ML and the MAP estimates and identifiable a.e. in m.s. by the LS estimate on the set if condition* (c6.1) *is satisfied.*

Proof. The assertion follows from Lemma 6.1 and Theorem 4.6. □

Now, consider the case, treated in Chapter 5, where the true system, given by Eq. (5.1), is assumed to belong to the set M_1 given by Eq. (5.2). Under conditions (c5.1) and (5.2), condition (6.1) simplifies to the following condition:

(c6.2) For each $j \in k; j \neq k$,

$$\left| \sum_j - \sum_k \right| \neq 0$$

Suppose that $k \in K$ is the true parameter. We have, for each $j \in K$; $j \neq k$,

$$\begin{aligned} I_n(k;j) &= -\frac{1}{2}\log\left|\sum_j\right| - \frac{1}{2}\mathrm{tr}\sum_j^{-1} E_*^{U_{n-1}}\{(z_n - \hat{z}_{j,n})(z_n - \hat{z}_{j,n})^T\} \\ &\quad + \frac{1}{2}\log\left|\sum_k\right| + \mathrm{tr}\sum_k^{-1} E_*^{U_{n-1}}\{(z_n - \hat{z}_{k,n})(z_n - \hat{z}_{k,n})^T\} \end{aligned} \tag{6.22}$$

where, for each $j \in k$,

$$
\begin{aligned}
&E_*^{U_{n-1}}\{(z_n - \hat{z}_{j,n})(z_n - \hat{z}_{j,n})^T\} \\
&= E_*^{U_{n-1}}\{z_n z_n^T\} - \hat{z}_{*,n}\hat{z}_{j,n}^T - \hat{z}_{j,n}\hat{z}_{*,n}^T + \hat{z}_{j,n}\hat{z}_{j,n}^T \\
&= \sum_* +(\hat{z}_{*,n} - \hat{z}_{j,n})(\hat{z}_{*,n} - \hat{z}_{j,n})^T \qquad (6.23)
\end{aligned}
$$

and, since $k = *$,

$$
\begin{aligned}
J_n(k;j) &\equiv \log \mathrm{h}_j^k(z_n|Z^{n-1}) - I_n(k;j) \\
&= \frac{1}{2}\mathrm{tr}\sum_k \left(\sum_j^{-1} - I\right) + \frac{1}{2}\mathrm{tr}\sum_k^{-1}[(z_n - \hat{z}_{k,n})](z_n - \hat{z}_{k,n})^T \\
&\quad - \frac{1}{2}\mathrm{tr}\sum_j^{-1}[(z_n - \hat{z}_{j,n})(z_n - \hat{z}_{j,n}) \\
&\quad - (\hat{z}_{k,n} - \hat{z}_{j,n})(\hat{z}_{k,n} - \hat{z}_{j,n})]^T \qquad (6.24)
\end{aligned}
$$

Since the sequences $(z_n - \hat{z}_{j,n})$ and $(\hat{z}_{k,n} - \hat{z}_{j,n})$ are ergodic for all $j, k \in K$, so are the sequences $(I_n(k;j))$ and $(J_n(k;j))$. It follows from Lemma 4.2 that condition (c4.2) is satisfied. Condition (c4.3) is satisfied if condition (5.4) is satisfied by Theorem 3.1 and the ergodicity of $(I_n(k;j))$. The identifiability of the system under condition (c5.4) thus follows from Theorem 4.5.

6.3 L_1 Convergence

We have shown in Section 4.4 that by bounding the information in the observations away from zero, bounds on the L_1 convergence rates of the likelihood and the *a posteriori* probability ratios can be established, which, in turn, provide performance measures for the ML and MAP estimation procedures. In this section, we consider the identification of a general class of time-varying systems driven by stochastic and deterministic inputs. The fact that only convergence in L_1 and not in the stronger senses of a.e. and m.s. is sought enable us to obtain rather explicit results. The stochastic and deterministic parts of the input are shown to contribute separately to the convergence rates of the identification procedures.

Consider the system

$$
\begin{aligned}
x_{n+1} &= F_{*,n}x_n + G_{*,n}u_n + \mathrm{J}_{*,n}w_n \\
z_n &= H_{*,n}x_n + v_n
\end{aligned} \tag{6.25}
$$

where (u_n) is a deterministic (known) input sequence and the other elements are as specified in Section 6.1. Also, consider a model set

$$
M_3 \equiv \{F_{j.n}, G_{j,n}, J_{j,n}, H_{j,n,}, \Psi_{j,n}, Q_{j,n}, R_{j,n}; j \in K\} \tag{6.26}
$$

where $Q_{j,n}$ and $\mathrm{R}_{j,n}$ are the covariance matrices of (u_n) and (v_n), respectively, corresponding to each model.

The incremental information for favoring a parameter k over a parameter j in the set K is given by Eq. (6.8). For each $j \in K$, we have

$$
I_n''(*;j) \equiv \frac{1}{2}(\hat{z}_{*,n} - \hat{z}_{j,n})^T \sum_{j,n}^{-1}(\hat{z}_{*,n} - \hat{z}_{j,n}) \tag{6.27}
$$

Let

$$
\hat{z}_{j,n}^* \equiv (\hat{z}_{*,n} - \hat{z}_{j,n}) = H_{J,N}^* \hat{x}_{j,n}^* \tag{6.28}
$$

where

$$
H_{j,n}^* \equiv (H_{*,n}, H_{j,n}) \tag{6.29}
$$

and

$$
\hat{x}_{j,n}^* \equiv (\hat{x}_{*,n}^T, \hat{x}_{j,n}^T)^T \tag{6.30}
$$

For each $j \in K'$, we have

$$
\begin{aligned}
\hat{x}_{j,n+1} = F_{j,n}(I - K_{j,n}H_{j,n})\hat{x}_{j,n} + G_{j,n}u_n \\
+ F_{j,n}K_{j,n}H_{*,n}x_n + F_{j,n}K_{j,n}v_n
\end{aligned} \tag{6.31}
$$

where

$$
K_{j,n} \equiv \sum_{j,n} H_{j,n}^T (H_{j,n} \sum_{j,n} H_{j,n}^T + R_{j,n})^{-1}
$$

Let

$$
\begin{aligned}
\bar{\hat{x}}_{j,n+1} \equiv E_* \hat{x}_{j,n+1} = F_{j,n}(I - K_{j,n}H_{j,n})\bar{\hat{x}}_{j,n} + G_{j,n}u_n \\
+ F_{j,n}K_{j,n}H_{*,n}\overline{x}_{*,n}
\end{aligned}
$$

where

$$\overline{x}_{*,n} \equiv E_* x_n = E_* E_*^{U_{n-1}} x_n = E_* \hat{x}_{*,n} = \overline{\hat{x}}_{*,n}$$

Also, let

$$\begin{aligned}\tilde{\hat{x}}_{j,n+1} \equiv \hat{x}_{j,n+1} - \overline{\hat{x}} &= F_{j,n}(I - K_{j,n}H_{j,n})\tilde{\hat{x}}_{j,n} \\ &+ F_{j,n}K_{j,n}H_{*,n}(x_n - \overline{x}_{*,n}) + F_{j,n}K_{j,n}v_n\end{aligned}$$

and

$$\tilde{x}_{*,n+1} \equiv x_{n+1} - \overline{x}_{*,n+1} = F_{*,n}\tilde{x}_{*,n} + J_{*,n}w_n$$

Now, let

$$\tilde{\hat{z}}^*_{j,n} \equiv H^*_{j,n}\tilde{\hat{x}}^*_{j,n} \tag{6.32}$$

where

$$\tilde{\hat{x}}^*_{j,n} \equiv (\tilde{\hat{x}}^T_{*,n}, \tilde{\hat{x}}^T_{j.n})^T \tag{6.33}$$

and let

$$\tilde{\hat{z}}^*_{j,n} \equiv H^*_{j,n}\overline{\hat{x}}^*_{j,n} \tag{6.34}$$

where

$$\overline{\hat{x}}^*_{j,n} \equiv (\overline{\hat{x}}^T_{*,n}, \overline{\hat{x}}^T_{j,n})^T \tag{6.35}$$

Then we can write

$$\begin{aligned}I''_n(*;j) &= \frac{1}{2}(\tilde{\hat{z}}^*_{j,n} + \overline{\hat{z}}^*_{j,n})^T \sum_{j,n}^{-1}(\tilde{\hat{z}}^*_{j,n} + \overline{\hat{z}}^*_{j,n}) \\ &= \frac{1}{2}\tilde{\hat{z}}^{*T}_{j,n}\sum_{j,n}^{-1}\tilde{\hat{z}}^*_{j,n} + \frac{1}{2}\overline{\hat{z}}^{*T}_{j,n}\sum_{j,n}^{-1}\overline{\hat{z}}^*_{j,n} + \tilde{\hat{z}}^{*T}_{j,n}\sum_{j,n}^{-1}\overline{\hat{z}}^*_{j,n}\end{aligned}$$

Let

$$\Phi^*_{j,n} \equiv \tilde{\hat{z}}^*_{j,n}\tilde{\hat{z}}^{*T}_{j,n} \tag{6.36}$$

$$\nabla^*_{j,n} \equiv \overline{\hat{z}}^*_{j,n}\overline{\hat{z}}^{*T}_{j,n} \tag{6.37}$$

$$\Theta^*_{j,n} \equiv \tilde{\hat{z}}^*_{j,n}\overline{\hat{z}}^{*T}_{j,n} \tag{6.38}$$

Then we have

$$I''_n(*;j) = \frac{1}{2}\mathrm{tr}\sum_{j,n}^{-1}(\Phi^*_{j,n} + \nabla^*_{j,n} + \Theta^*_{j,n}) \tag{6.39}$$

We shall use

$$\overline{\Phi}^*_{j,n} \equiv E_* \Phi^*_{j,n}$$

which is obtained via the following procedure.

Define

$$\tilde{x}^*_{j,n} \equiv (\tilde{x}_{*,n}, \tilde{\hat{x}}_{*,n}, \tilde{x}_{j,n}) \tag{6.40}$$

Then $\tilde{x}^*_{j,n}$ is generated by the following equation:

$$\tilde{x}^*_{j,n+1} = F^*_{j,n} \tilde{x}^*_{j,n} + G^*_j w_n \tag{6.41}$$

where

$$F^*_{j,n} = \begin{bmatrix} F_{*,n} & 0 & 0 \\ F_{*,n} K_{*,n} H_{*,n} & F_{*,n}(I - K_{*,n} H_{*,n}) & 0 \\ F_{j,n} K_{j,n} H_{*,n} & 0 & F_{j,n}(I - K_{j,n} H_{h,n}) \end{bmatrix} \tag{6.42}$$

$$G^*_{j,n} \equiv \begin{bmatrix} J_{*,n} & 0 \\ 0 & F_{*,n} K_{*,n} \\ 0 & F_{j,n} K_{j,n} \end{bmatrix}; \quad W_n \equiv \begin{bmatrix} w_n \\ v_n \end{bmatrix} \tag{6.43}$$

Also, let

$$Q^*_n \equiv \begin{bmatrix} Q_{*,n} & 0 \\ 0 & R_{*,n} \end{bmatrix} \tag{6.44}$$

Then

$$\overline{\Phi}^*_{j,n} = (\underline{0}, \mathrm{H}^*_{j,n}) \Pi^*_{j,n} (\underline{0}, \mathrm{H}^*_{j,n})^T \tag{6.45}$$

where

$$\Pi^*_{j,n} \equiv E_* \{ \tilde{x}^*_{j,n} \tilde{x}^{*^T}_{j,n} \} \tag{6.46}$$

is generated by the equation

$$\Pi^*_{j,n+1} = F^*_{j,n} \Pi^*_{j,n} F^{*^T}_{j,n} + G^*_{j,n} Q^*_n G^{*^T}_{j,n} \tag{6.47}$$

initialized at

$$\Pi_j^* = \begin{bmatrix} \Psi_* & 0 & 0 \\ 0 & \Psi_* & 0 \\ 0 & 0 & \Psi_j \end{bmatrix}$$

Next, consider $\nabla_{j,n}^*$. We have

$$\overline{x}_{j,n+1}^* = \overline{F}_{j,n}^* \overline{x}_{j,n}^* + \overline{G}_{j,n}^* u_n; \quad \overline{x}_{j,0}^* = 0 \tag{6.48}$$

where

$$\overline{F}_{j,n}^* \equiv \begin{bmatrix} F_{*,n} & 0 \\ F_{j,n} K_{j,n} H_{*,n} & F_{j,n}(I - K_{j,n} H_{j,n}) \end{bmatrix}; \quad \overline{G}_{j,n}^* \equiv \begin{bmatrix} G_{*,n} \\ G_{j,n} \end{bmatrix} \tag{6.49}$$

Equation (6.48) can be written as

$$\bar{x}_{j,n}^* = \sum_{m=1}^{n-1} A_j^*(n,m) \overline{G}_{j,m}^* u_m \tag{6.50}$$

where

$$A_j^*(n,m) \equiv \Pi_{i=1}^{n-m-1} F_{j,i} \tag{6.51}$$

We have

$$E_* \Theta_{j,n} = 0$$

Thus, for $k = *$ and each $j \in K$; $j \neq k$

$$\begin{aligned} \overline{I}_n(k;j) &= \overline{I}'_n(k;j) + \overline{I}''_n(k;j) \\ &= \frac{1}{2} \log \frac{|\sum_{j,n}|}{|\sum_{k,n}|} + \frac{1}{2} \mathrm{tr} \left(\sum_{k,n} \sum_{j,n}^{-1} - I \right) + \frac{1}{2} \mathrm{tr} \sum_{j,n}^{-1} (\overline{\Phi}_{j,n}^k + \nabla_{j,n}^k) \end{aligned}$$

Note that, while $\sum_{j,n}$ and $\sum_{k,n}$(or $I'_n(k:j)$) depend only on the stochastic part of the input, the term $\nabla_{j,n}^k$ represents its deterministic part. In the sequel, we examine the separate contributions of these elements to the L_1 convergence rates of the likelihood ratios and the *a posteriori* probability ratios on the set K.

Theorem 6.2. *Suppose that the true system in Eq. (6.25) belongs to the set M_3 given by Eq. (6.26). Let $k \in K$ be the true parameter. Suppose that for each $j \in K; j \neq k$ there exists a positive scalar α_j and a positive integer N_j such that*

$$\left\| \sum_{k,n} - \sum_{j,n} \right\| \geq \alpha_j \quad \text{for all } n \geq N_j \tag{6.52}$$

Then we have, for each $j \in K; j \neq k$,

$$E_* h_j^k(Z^n) \geq e^{(n-N_j+1)\alpha_j} \quad \text{for all } n \geq N_j$$

and

$$E_* \frac{f^b(k|Z^n)}{f^b(j|Z^n)} = \frac{f^b(k)}{f^b(j)} e^{(n-N_J+1)\alpha_J} \quad \text{for all } n \geq N_j$$

Proof. Following arguments similar to those made in the proof of Lemma 6.1, we first show that Eq. (6.52) implies that there exists some $\varepsilon > 0$ such that

$$|\lambda_n - 1| \geq \varepsilon \quad \text{for each } i = 1, \ldots, \ell, \quad \text{for all } n \geq N_j \tag{6.53}$$

where $\lambda_{n,i}; i = 1, \ldots, \ell$ are the solutions of Eq. (6.13). Suppose that Eq. (6.53) does not hold, then for any $\varepsilon > 0$, there exists some $n_\varepsilon \geq N_j$ such that

$$|\lambda_{n_\varepsilon} - 1_{n_\varepsilon}| < \varepsilon \tag{6.54}$$

where

$$\lambda_{n_\varepsilon, i_{n_\varepsilon}} \equiv \min\{\lambda_{n_\varepsilon,i};\ i = 1, \ldots, \ell\}$$

and then, by continuity of the left-hand side of Eq. (6.13), in λ_n, given $\alpha_j > 0$, one can take ε such that

$$\left\| \sum_{k,n_\varepsilon} - \lambda_{n,i_{n_\varepsilon}} \sum_{j,n} | - | \sum_{k,n} - \sum_{j,n} \right\| < \alpha_j$$

yielding

$$\left\| \sum_{k,n} - \sum_{j,n} \right\| < \alpha_j$$

contradicting Eq. (6.52). Hence, Eqs. (6.52) implies (6.53). Now, by Eq. (6.16) and by the convexity of Eq. (6.20), we have that Eq. (6.53) implies that for each $j \in K; j \notin k$, there exists some α_j such that

$$\overline{I}'_n(k;j) \geq \alpha_j \quad \text{for all } n \geq N_j$$

Since $\overline{I}''_n(k;j) \geq 0$, we then have

$$\overline{I}_n(k;j) \geq \alpha_j \quad \text{for all } n \geq N_j$$

Condition (c4.5) is then satisfied, and the assertion follows from Eqs. (4.11) and (4.12) in the proof of Theorem 4.11. □

Corollary 6.1. *Let the set M_2 be time-invariant and let the true system belong to M_2. Suppose that for each $j \in K, \sum_j$ given by Theorem* 6.1 *is finite and nonsingular. Then the L_1-convergence bounds asserted in Theorem* 6.2 *hold under condition* (c6.2), *where k is the true parameter.*

Proof. Condition (c6.2) implies that for each $j \in K; j \neq k$, there exists some $\varsigma_j > 0$ such that

$$\left\| \sum_j - \sum_k \right\| \geq \varsigma_j$$

Clearly,

$$\lim_{n\to\infty} \left\| \sum_{j.n} - \sum_{k,n} \right\| = \left\| \sum_j - \sum_k \right\| \geq \varsigma_j$$

Hence, for any positive scalar α_j such that $0 < \alpha_j < \varsigma_j$, there exists some positive integer N_j such that

$$\left\| \sum_{j,n} - \sum_{k,n} \right\| \geq \alpha_j \quad \text{for all } n \geq N_j$$

The assertion then follows from Theorem 6.2. □

Theorem 6.3. *Suppose that the true system in Eq. (6.25) belongs to the set M_3 given by Eq. (6.26). Let $k \in K$ be the true parameter. Suppose that for each $j \in K; j \notin k$ there exists a positive scalar α_j and a positive integer N_j such that*

$$\operatorname{tr} \sum_{j,n}^{-1} \nabla_{j,n}^{k} \geq 2\alpha_j \quad \text{for all } n \geq N_j \tag{6.55}$$

Then the L_1 convergence rates asserted in Theorem 6.2 hold.

Proof. For each $j \in K; j \notin k$, we have

$$I'_n(k;j) \geq \alpha_j \quad \text{for all } n \geq 0$$

(it follows from Eq. (3.5)). Also, see the proof of Lemma 6.1 and

$$\operatorname{tr} \sum_{j,n}^{-1} \phi_{j,n}^{k} \geq 0 \quad \text{for all } n \geq 0$$

Hence,

$$\overline{I}_n(k;j) \geq \frac{1}{2} \operatorname{tr} \sum_{j,n}^{-1} \nabla_{j,n}^{k} \geq \alpha_j \quad \text{for all } n \geq N_j$$

Condition (c4.5) is then satisfied, and the assertion follows from Eqs. (4.11) and (4.12) in the proof of Theorem 4.11. □

Theorem 6.2 guarantees a certain L_1 convergence rate of the likelihood ratios and the *a posteriori* probability ratios under a certain condition involving the stochastic characteristics of the inputs to the system. Theorem 6.3 means that the convergence rates can be improved by the application of certain deterministic inputs, satisfying Eq. (6.55).

Chapter 7

Firing-Rate Dynamics Irreversibility

7.1 Introduction

Irreversibility of natural processes is implied by the second law of thermodynamics (Mandl, 1988). In formal terms, the mathematical models of such processes are noninvertible. Here, we show that the widely recognized, experimentally supported model of neural firing-rate dynamics is noninvertible and, hence, time-irreversible. This fundamental inability to reveal past values of a firing rate process from present values would seem to present an inherent behavioral difficulty, as present behavior should somehow employ past experience. Luckily, as stressed in a previous book (Baram, 2021), memories of past experience are encapsulated in the form of global dynamic attractors rather than actual values of firing rates. The noninvertibility of the firing-rate process motivates, then, the central role played by the singularities of the model as a language of cortical dynamics. To this end, we make room for a detailed proof of the noninvertibility of the model, as revealed by an earlier work (Baram, 2012).

7.2 Discrete Iteration Map of Neural Firing and Synaptic Plasticity

The vector-valued truncated discrete-time version of the spike response model, Eq. (2.16), for the firing rates is given by

$$\boldsymbol{\upsilon}(k+1) = \mathbf{K}_\upsilon \boldsymbol{\upsilon}(k) + \mathbf{T}_\upsilon^{-1}\mathbf{f}\left(\mathbf{W}(k)^T \sum_{p=0}^{N-1} \mathbf{E}_p^{-1}\boldsymbol{\upsilon}(k-p) + \mathbf{u}(k)\right) \tag{7.1}$$

where $\boldsymbol{\upsilon}(k)$ is the vector of firing rates corresponding to the neurons at time instance k, $\mathbf{W}(k)$ is the matrix of synaptic weights, $\mathbf{u}(k)$ is the vector of conductance-based activation inputs, $\mathbf{K}_\upsilon = \text{diag}[(\tau_{\upsilon_i} - 1)/\tau_{\upsilon_i}]_{i=1,\ldots,n}$, $\mathbf{T}_{\boldsymbol{\upsilon}} = \text{diag}[\tau_{\upsilon_i}]_{i=1,\ldots,n}$ and $\mathbf{E}_p = \text{diag}\,[e^{p/\tau_i}]_{i=1,\ldots,n}$ are diagonal matrices of the corresponding elements, $\mathbf{f}(\mathbf{x}) = \text{vec}[f(x_i)]_{i=1,\ldots,n}$ and $\mathbf{u}(k) = \text{vec}[u_i(k)]_{i=1,\ldots,n}$ are vectors of the corresponding elements, and $\mathbf{W}(k) = [\boldsymbol{\omega}_1(k)\boldsymbol{\omega}_2(k)\cdots\boldsymbol{\omega}_n(k)]$ is a matrix whose columns are $\boldsymbol{\omega}_i(k), i = 1,\ldots,n$. N is a positive integer which satisfies the desired relative error ρ_i in Eq. (2.24) for any i, and $\mathbf{f}$ is a vector-valued function, whose scalar elements are defined by f_i of Eq. (2.17).

A discrete-time matrix-valued version of Eq. (2.18) is given by

$$\begin{aligned}\mathbf{W}(k+1) = \mathbf{W}(k) + \mathbf{T}_\omega^{-1}\sum_{p=0}^{N-1}\mathbf{E}_p^{-1}\big[\big[\boldsymbol{\upsilon}(k-p)\\ - \boldsymbol{\theta}(k)\big]\boldsymbol{\upsilon}^2(k-p)^T\big] * \mathbf{S}\end{aligned} \tag{7.2}$$

with

$$\boldsymbol{\theta}(k) = \mathbf{T}_\theta^{-1}\sum_{p=0}^{N-1}\mathbf{E}_p^{-1}\boldsymbol{\upsilon}^2(k-p) \tag{7.3}$$

where $\mathbf{T}_\omega = \text{diag}[\tau_{\omega_i}]_{i=1,\ldots,n}$, $\mathbf{T}_\theta = \text{diag}[\tau_{\theta_i}]_{i=1,\ldots,n}$ and $\mathbf{S} = [\mathbf{s}_1\mathbf{s}_2\ldots\mathbf{s}_n]$, $\boldsymbol{\upsilon}^2(k-p)$ is the vector whose ith element is $\upsilon_i^2(k-p)$, $\mathbf{S}$ is the connectivity matrix, having the value 1 in all positions corresponding to connected neurons and the value 0 elsewhere, and

* denotes the elementwise product between two matrices of the same dimension, as in

$$\mathbf{A} * \mathbf{X} = \begin{bmatrix} a_{1,1}x_{1,1} & a_{1,2}x_{1,2} & \cdots & a_{1,r}x_{1,r} \\ a_{2,1}x_{2,1} & a_{2,2}x_{2,2} & \cdots & a_{2,r}x_{2,r} \\ \vdots & & & \\ a_{p,1}x_{p,1} & a_{p,2}x_{p,2} & \cdots & a_{p,r}x_{p,r} \end{bmatrix} \tag{7.4}$$

with $a_{i,j}$ and $x_{i,j}$ as the respective scalar elements.

7.3 Noninvertibility (Time-Irreversibility) of the Map

The map, Eqs. (7.1)–(7.3), is invertible if and only if it has an inverse which yields, for any values of the map parameters and any real values of the firing rates at time $k+1$, unique real values for the firing rates at time k. If this is not the case, the map is said to be noninvertible. It should be noted that only positive (including zero) values are acceptable for the firing rates.

Theorem 7.1. *The map, Eqs.* (7.1)–(7.3), *is noninvertible.*

Proof. Denoting

$$\boldsymbol{\upsilon}_p(k) = \boldsymbol{\upsilon}(k-p) \tag{7.5}$$

Eq. (7.1) can be written as

$$\boldsymbol{\upsilon}_0(k+1) = \mathbf{K}_\upsilon \boldsymbol{\upsilon}_0(k) + \mathbf{T}_\upsilon^{-1}\mathbf{f}\left(\mathbf{W}(k)^T \sum_{p=0}^{N-1} \mathbf{E}_p^{-1}\boldsymbol{\upsilon}_p(k) + \mathbf{u}(k)\right) \tag{7.6}$$

and Eq. (7.2) can be written as

$$\begin{aligned}\mathbf{W}(k+1) = \mathbf{W}(k) &+ \mathbf{T}_\omega^{-1} \sum_{p=0}^{N-1} \mathbf{E}_p^{-1}[\boldsymbol{\upsilon}_p(k)\boldsymbol{\upsilon}_p^2(k)^T] * \mathbf{S} \\ &- \mathbf{T}_\omega^{-1} \sum_{p=0}^{N-1} \mathbf{E}_p^{-1}[\boldsymbol{\theta}(k)\boldsymbol{\upsilon}_p^2(k)^T] * \mathbf{S}\end{aligned} \tag{7.7}$$

where, by Eq. (7.3),

$$\boldsymbol{\theta}(k) = \mathbf{T}_\theta^{-1} \sum_{p=0}^{N-1} \mathbf{E}_p^{-1} \boldsymbol{\upsilon}_p^2(k) \tag{7.8}$$

Substituting Eqs. (7.7) and (7.8) into Eq. (7.6) and employing Eq. (7.5), we obtain

$$\begin{aligned}
\boldsymbol{\upsilon}_0(k+1) &= \mathbf{K}_\upsilon \boldsymbol{\upsilon}_0(k) + \mathbf{T}_\upsilon^{-1}\mathbf{f}(\mathbf{P}(\boldsymbol{\upsilon}_p(k), p = 0, \ldots, N-1)) \\
\boldsymbol{\upsilon}_1(k+1) &= \boldsymbol{\upsilon}_0(k) \\
\boldsymbol{\upsilon}_2(k+1) &= \boldsymbol{\upsilon}_1(k) \\
&\vdots \\
\boldsymbol{\upsilon}_{N-1}(k+1) &= \boldsymbol{\upsilon}_{N-2}(k)
\end{aligned} \tag{7.9}$$

where

$$\begin{aligned}
&\mathbf{P}(\boldsymbol{\upsilon}_p(k), p = 0, \ldots, N-1) \\
&= -\left[\sum_{p=0}^{N-1} \mathbf{S}^T * \left[\boldsymbol{\upsilon}_p^2(k) \sum_{p=0}^{N-1} \boldsymbol{\upsilon}_p^2(k)^T \mathbf{E}_p^{-1}\right] \mathbf{T}_\theta^{-1} \mathbf{E}_p^{-1}\right] \\
&\quad \times \mathbf{T}_\omega^{-1} \sum_{p=0}^{N-1} \mathbf{E}_p^{-1} \boldsymbol{\upsilon}_p(k) + \left[\sum_{p=0}^{N-1} \mathbf{S}^T * [\boldsymbol{\upsilon}_p(k) \boldsymbol{\upsilon}_p^2(k)^T] \mathbf{E}_p^{-1}\right] \\
&\quad \times \mathbf{T}_\omega^{-1} \sum_{p=0}^{N-1} \mathbf{E}_p^{-1} \boldsymbol{\upsilon}_p(k) + \mathbf{W}(k+1)^T \sum_{p=0}^{N-1} \mathbf{E}_p^{-1} \boldsymbol{\upsilon}_p(k) + \mathbf{u}(k)
\end{aligned} \tag{7.10}$$

Given a unique positive real value for $\boldsymbol{\upsilon}_i(k+1), i = 1, \ldots, N-1$, the last $N-1$ equations of Eq. (7.9) yield unique positive real values for $\boldsymbol{\upsilon}_i(k), i = 0, \ldots, N-2$. It remains to be determined whether, given, in addition, $\boldsymbol{\upsilon}_0(k+1)$, Eq. (7.9) provides a unique positive real value for $\boldsymbol{\upsilon}_{N-1}(k)$. The threshold nonlinearity in Eq. (2.17) and the first equation of Eq. (7.9) imply that either

$$\mathbf{P}(\boldsymbol{\upsilon}_p(k), p = 0, \ldots, N-1) < 0 \tag{7.11}$$

or

$$\boldsymbol{v}_0(k+1) = \mathbf{K}_\upsilon \boldsymbol{v}_0(k) + \mathbf{T}_\upsilon^{-1}(\mathbf{P}(\boldsymbol{v}_p(k), p = 0, \ldots, N-1)) \quad (7.12)$$

If Eq. (7.11) is satisfied, then, by the first equation of Eq. (7.9),

$$\boldsymbol{v}_0(k+1) = \mathbf{K}_\upsilon \boldsymbol{v}_1(k+1) \quad (7.13)$$

Hence, Eq. (7.12) is satisfied in the range

$$\boldsymbol{v}_0(k+1) \neq \mathbf{K}_\upsilon \boldsymbol{v}_1(k+1) \quad (7.14)$$

In order to show that the map is noninvertible, it will suffice, then, to show that Eq. (7.12) does not have a unique real-valued solution for $\boldsymbol{v}_{N-1}(k)$. Substituting Eq. (7.10) into Eq. (7.12), we have

$$\begin{aligned}
&\mathbf{T}_\upsilon^{-1}\left[\sum_{p=0}^{N-1} \mathbf{S}^T * \left[\boldsymbol{v}_p^2(k) \sum_{p=0}^{N-1} \boldsymbol{v}_p^2(k)^T \mathbf{E}_p^{-1}\right] \mathbf{T}_\theta^{-1} \mathbf{E}_p^{-1}\right] \mathbf{T}_\omega^{-1} \sum_{p=0}^{N-1} \mathbf{E}_p^{-1} \boldsymbol{v}_p(k) \\
&\quad - \mathbf{T}_\upsilon^{-1}\left[\sum_{p=0}^{N-1} \mathbf{S}^T * [\boldsymbol{v}_p(k) \boldsymbol{v}_p^2(k)^T] \mathbf{E}_p^{-1}\right] \mathbf{T}_\omega^{-1} \sum_{p=0}^{N-1} \mathbf{E}_p^{-1} \boldsymbol{v}_p(k) \\
&\quad - \mathbf{K}_\upsilon \boldsymbol{v}_0(k) - \mathbf{T}_\upsilon^{-1} \mathbf{W}(k+1)^T \sum_{p=0}^{N-1} \mathbf{E}_p^{-1} \boldsymbol{v}_p(k) \\
&\quad - \mathbf{T}_\upsilon^{-1} \boldsymbol{\beta}(k) - \boldsymbol{v}_0(k+1) = 0 \qquad (7.15)
\end{aligned}$$

The left-hand side of Eq. (7.15) can be written as a polynomial in $\boldsymbol{v}_{N-1}(k)$:

$$\begin{aligned}
P(\boldsymbol{v}_{N-1}(k)) &= P_5(\boldsymbol{v}_{N-1}(k)) + P_4(\boldsymbol{v}_{N-1}(k)) + P_4(\boldsymbol{v}_{N-1}(k)) \\
&\quad + P_3(\boldsymbol{v}_{N-1}(k)) + P_1(\boldsymbol{v}_{N-1}(k)) \\
&\quad + P_0(\boldsymbol{v}_{N-1}(k)) \qquad (7.16)
\end{aligned}$$

where, incorporating Eq. (7.9),

$$\begin{aligned}
&P_5(\boldsymbol{v}_{N-1}(k)) \\
&\quad = \mathbf{T}_\upsilon^{-1}[\mathbf{S}^T * [\boldsymbol{v}_{N-1}^2(k) \boldsymbol{v}_{N-1}^2(k)^T]] \\
&\qquad \times \mathbf{E}_{N-1}^{-1} \mathbf{T}_\theta^{-1} \mathbf{E}_{N-1}^{-1} \mathbf{T}_\omega^{-1} \mathbf{E}_{N-1}^{-1} \boldsymbol{v}_{N-1}(k)
\end{aligned}$$

$$
\begin{aligned}
&P_4(\boldsymbol{\upsilon}_{N-1}(k)) \\
&\quad = T_\upsilon^{-1}[\mathbf{S}^T * [\boldsymbol{\upsilon}_{N-1}^2(k)\boldsymbol{\upsilon}_{N-1}^2(k)^T]] \\
&\qquad \times \mathbf{E}_{N-1}^{-1}\mathbf{T}_\theta^{-1}\mathbf{E}_{N-1}^{-1}\mathbf{T}_\omega^{-1}\sum_{p=1}^{N-1}\mathbf{E}_p^{-1}\boldsymbol{\upsilon}_p(k+1) \\
&P_3(\boldsymbol{\upsilon}_{N-1}(k)) \\
&\quad = -\mathbf{T}_\upsilon^{-1}[\mathbf{S}^T * [\boldsymbol{\upsilon}_{N-1}^2(k)\boldsymbol{\upsilon}_{N-1}^2(k)^T]]\mathbf{E}_{N-1}^{-1}\mathbf{T}_\omega^{-1}\mathbf{E}_{N-1}^{-1}\boldsymbol{\upsilon}_{N-1}(k) \\
&P_2(\boldsymbol{\upsilon}_{N-1}(k)) \\
&\quad = -\mathbf{T}_\upsilon^{-1}[\mathbf{S}^T * [\boldsymbol{\upsilon}_{N-1}^2(k)\boldsymbol{\upsilon}_{N-1}^2(k)^T]] \\
&\qquad \times \mathbf{E}_{N-1}^{-1}\mathbf{T}_\omega^{-1}\mathbf{E}_{N-1}^{-1}\sum_{p=1}^{N-1}\mathbf{E}_p^{-1}\boldsymbol{\upsilon}_p(k+1) \\
&P_1(\boldsymbol{\upsilon}_{N-1}(k)) \\
&\quad = \mathbf{T}_\upsilon^{-1}\boldsymbol{\omega}(k+1)^T\mathbf{E}_{N-1}^{-1}\boldsymbol{\upsilon}_{N-1}(k) \\
&P_0(\boldsymbol{\upsilon}_{N-1}(k)) \\
&\quad = \mathbf{T}_\upsilon^{-1}\left[\sum_{p=1}^{N-1}\mathbf{S}^T * [\boldsymbol{\upsilon}_{N-1}^2(k+1)\sum_{p=1}^{N-1}\boldsymbol{\upsilon}_p^2(k+1)^T\mathbf{E}_p^{-1}]\mathbf{T}_\theta^{-1}\mathbf{E}_p^{-1}\right] \\
&\qquad \times \mathbf{T}_\omega^{-1}\sum_{p=1}^{N-1}\mathbf{E}_p^{-1}\boldsymbol{\upsilon}_p(k+1) \\
&\qquad - \mathbf{T}_\upsilon^{-1}\left[\sum_{p=1}^{N-1}\mathbf{S}^T * [\boldsymbol{\upsilon}_p(k+1)\boldsymbol{\upsilon}_p^2(k+1)^T]\mathbf{E}_p^{-1}\right] \\
&\qquad \times \mathbf{T}_\omega^{-1}\sum_{p=1}^{N-1}\mathbf{E}_p^{-1}\boldsymbol{\upsilon}_p(k+1) \\
&\qquad + \mathbf{T}_\upsilon^{-1}\left[\mathbf{W}(k+1)^T\sum_{p=1}^{N-1}\mathbf{E}_p^{-1}\boldsymbol{\upsilon}_p(k+1)\right] \\
&\qquad + \mathbf{K}_\upsilon\boldsymbol{\upsilon}_1(k+1) - \boldsymbol{\upsilon}_0(k+1) + \mathbf{T}_\upsilon^{-1}\mathbf{u}(k)
\end{aligned} \tag{7.17}
$$

The ith row of the vector in Eq. (7.15) is a scalar polynomial equation of order 5 in the elements of $\boldsymbol{\upsilon}_{i,N-1}(k)$. The terms of Eq. (7.17) corresponding to the scalar polynomial are, in descending order of power of $\boldsymbol{\upsilon}_{i,N-1}(k)$,

$$
\begin{aligned}
p_{i,5}(\boldsymbol{\upsilon}_{N-1}(k)) &= \frac{e^{-3(N-1)/\tau_i}}{\tau_{\upsilon_i}\tau_{\theta_i}\tau_{\omega_i}} \upsilon_{i,N-1}^2(k) \sum_{j=1}^{n} \upsilon_{j,N-1}^3(k) \\
p_{i,4}(\boldsymbol{\upsilon}_{N-1}(k)) &= \frac{e^{-2(N-1)/\tau_i}}{\tau_{\upsilon_i}\tau_{\theta_i}\tau_{\omega_i}} \upsilon_{i,N-1}^2(k) \\
&\quad \times \sum_{j=1}^{n} \left[\upsilon_{j,N-1}^2(k) \sum_{p=0}^{N-2} e^{-p/\tau_j} \upsilon_{j,p}(k) \right] \\
p_{i,3}(\boldsymbol{\upsilon}_{N-1}(k)) &= -\frac{e^{-2(N-1)/\tau_i}}{\tau_{\upsilon_i}\tau_{\omega_i}} \upsilon_{i,N-1}(k) \sum_{j=1}^{n} \upsilon_{j,N-1}^3(k) \\
p_{i,2}(\boldsymbol{\upsilon}_{N-1}(k)) &= -\frac{e^{-2(N-1)/\tau_i}}{\tau_{\upsilon_i}\tau_{\omega_i}} \upsilon_{i,N-1}(k) \\
&\quad \times \sum_{j=1}^{n} \left[\upsilon_{j,N-1}^2(k) \sum_{p=0}^{N-2} e^{-p/\tau_j} \upsilon_{j,p}(k) \right] \\
p_{i,1}(\boldsymbol{\upsilon}_{N-1}(k)) &= \frac{e^{-(N-1)/\tau_i}}{\tau_{\upsilon_i}} \sum_{j=1}^{n} \omega_{i,j}(k+1) \upsilon_{j,N-1}(k) \\
p_{i,0}(\boldsymbol{\upsilon}_{N-1}(k)) &= [\boldsymbol{\mu}(k+1)]_i + [\mathbf{T}_{\upsilon}^{-1}\mathbf{u}(k)]_i
\end{aligned}
\tag{7.18}
$$

where $\omega_{i,j}(k+1)$ is the i, jth scalar element of $\mathbf{W}(k+1)$, and $[\mathbf{T}_{\upsilon}^{-1}\mathbf{u}(k)]_i$ and $[\boldsymbol{\mu}(k+1)]_i$ are the ith scalar elements of the vectors $\mathbf{T}_{\upsilon}^{-1}\mathbf{u}(k)$ and

$$
\begin{aligned}
\boldsymbol{\mu}(\mathbf{v}(k+1)) = \mathbf{T}_{\upsilon}^{-1} & \left[\sum_{p=1}^{N-1} \mathbf{S}^T * \left[\boldsymbol{\upsilon}_p^2(k+1) \sum_{p=1}^{N-1} \boldsymbol{\upsilon}_p^2(k+1)^T \mathbf{E}_p^{-1} \right] \right. \\
& \left. \times \ \mathbf{T}_{\theta}^{-1}\mathbf{E}_p^{-1} \right] \mathbf{T}_{\omega}^{-1} \sum_{p=1}^{N-1} \mathbf{E}_p^{-1} \boldsymbol{\upsilon}_p(k+1)
\end{aligned}
$$

$$
\begin{aligned}
&- \mathbf{T}_{\upsilon}^{-1}\left[\sum_{p=1}^{N-1} \mathbf{S}^{T} * [\boldsymbol{\upsilon}_p(k+1)\boldsymbol{\upsilon}_p^2(k+1)^T]\mathbf{E}_p^{-1}\right] \\
&\times \mathbf{T}_{\omega}^{-1}\sum_{p=1}^{N-1} \mathbf{E}_p^{-1}\boldsymbol{\upsilon}_p(k+1) \\
&- \mathbf{T}_{\upsilon}^{-1}\left[\mathbf{W}(k+1)^T \sum_{p=1}^{N-1} \mathbf{E}_p^{-1}\boldsymbol{\upsilon}_p(k+1)\right] \\
&- \mathbf{K}_{\upsilon}\boldsymbol{\upsilon}_1(k+1) = \boldsymbol{\upsilon}_0(k+1) \qquad (7.19)
\end{aligned}
$$

respectively. While the map at hand is polynomial in the firing rates, it is also a transcendental (nonalgebraic) function of the initial conditions and the parameter values (including, in our case, external activation inputs. Indeed, local bifurcations are caused by changes in these parameters. Reduction to circuits of identical neurons is justified by the Hebbian paradigm, which implies segregation of mutually asynchronous circuits or, as considered in the subsequent chapter, such segregation as enforced by synapse polarization.

If all the neurons have the same parameter values and the same initial conditions, they all must behave in the same manner. This means that within this parametric domain (the domain of symmetry), all the scalar rows of Eq. (7.15) represent the same scalar polynomial equation. If the ith neuron is self-connected, the ith row of Eq. (7.15) can be written as a univariate polynomial equation in $\upsilon_{i,N-1}(k)$, whose terms are, in descending order of power,

$$p_{i,5}(\upsilon_{i,N-1}(k)) = \frac{ne^{-3(N-1)/\tau_i}}{\tau_{\upsilon_i}\tau_{\theta_i}\tau_{\omega_i}}\upsilon_{i,N-1}^5(k)$$

$$p_{i,4}(\upsilon_{i,N-1}(k)) = \frac{ne^{-2(N-1)/\tau_i}\sum_{p=0}^{N-2} e^{-p/r_j}\upsilon_{i,p}(k)}{\tau_{\upsilon_i}\tau_{\theta_i}\tau_{\omega_i}}\upsilon_{i,N-1}^4(k)$$

$$p_{i,3}(\upsilon_{i,N-1}(k)) = -\frac{ne^{-2(N-1)/\tau_i}}{\tau_{\upsilon_i}\tau_{\omega_i}}\upsilon_{i,N-1}^3(k)$$

$$p_{i,2}(\upsilon_{i,N-1}(k)) = -\frac{ne^{-2(N-1)/\tau_i}\sum_{p=0}^{N-2} e^{-p/\tau_j}\upsilon_{i,p}(k)}{\tau_{\upsilon_i}\tau_{\omega_i}}\upsilon_{i,N-1}^2(k)$$

$$p_{i,1}(\upsilon_{i,N-1}(k)) = \frac{ne^{-(N-1)/\tau_i}\omega(k+1)}{\tau_{\upsilon_i}}\upsilon_{i,N-1}(k)$$
$$p_{i,0}(\upsilon_{i,N-1}(k)) = [\boldsymbol{\mu}(\mathbf{v}(k+1)) + \mathbf{T}_{\upsilon}^{-1}\mathbf{u}(k)]_i \tag{7.20}$$

(note that, due to the assumed parametric identity, all the synaptic weights are the same, hence $\omega(k+1) = \omega_{i,j}(k+1); i, j = 1, \ldots, n$). If the neurons are self-disconnected, n will be replaced by $n-1$ in all but the term for $p_{i,1}(\boldsymbol{\upsilon}_{N-1}(k))$. Descartes' rule of signs (Fine and Rosenberg, 1997) states that the number of real positive roots of a univariate polynomial equation with real coefficients equals the number of sign changes or is less than it by a multiple of two when the coefficients are ordered according to descending powers of the polynomial. It can be seen that, since $\boldsymbol{\upsilon}_{i,N-1}(k)$ is positive, there is an even number of Descartes sign changes in Eq. (7.20) within the domain:

$$[\boldsymbol{\mu}(\mathbf{v}(k+1)) + \mathbf{T}_{\upsilon}^{-1}\mathbf{u}(k)]_i > 0 \tag{7.21}$$

In order to see that this is true, let us write Eq. (7.16) for the scalar case as

$$\begin{aligned} p(\boldsymbol{\upsilon}_{N-1}(k)) = {} & p_{i,5}(\boldsymbol{\upsilon}_{N-1}(k)) + p_{i,4}(\boldsymbol{\upsilon}_{N-1}(k)) + p_{i,3}(\boldsymbol{\upsilon}_{N-1}(k)) \\ & + p_{i,2}(\boldsymbol{\upsilon}_{N-1}(k)) + p_{i,1}(\boldsymbol{\upsilon}_{N-1}(k)) \\ & + p_{i,0}(\boldsymbol{\upsilon}_{N-1}(k)) \end{aligned} \tag{7.22}$$

which can be rearranged in the order of decreasing powers of $\boldsymbol{\upsilon}_{N-1}(k)$ as

$$\begin{aligned} p(\boldsymbol{\upsilon}_{N-1}(k)) = {} & p_{i,5}(\boldsymbol{\upsilon}_{N-1}(k)) + [p_{i,4}(\boldsymbol{\upsilon}_{N-1}(k)) + p_{i,3}(\boldsymbol{\upsilon}_{N-1}(k))] \\ & + p_{i,2}(\boldsymbol{\upsilon}_{N-1}(k)) \\ & + p_{i,1}(\boldsymbol{\upsilon}_{N-1}(k)) + p_{i,0}(\boldsymbol{\upsilon}_{N-1}(k)) \end{aligned} \tag{7.23}$$

It can be seen that, according to Eq. (7.20), the right-hand side of Eq. (7.23) has two Descartes' sign changes regardless of the sign of $[p_{i,4}(\boldsymbol{\upsilon}_{N-1}(k)) + p_{i,3}(\boldsymbol{\upsilon}_{N-1}(k))]$.

It follows that, within the domain in Eq. (7.21), the map, under parametric identity, has no unique positive real inverse, hence it is noninvertible. This holds for networks of any number of self-connected neurons and for networks of two or more self-disconnected neurons (for a single self-disconnected neuron, all the terms of Eq. (7.20) but the last two vanish, and the map becomes invertible). □

7.4 Discussion

We have analytically shown that, depending on parameter values, the neural firing rate map is noninvertible. Noninvertibility of the map motivates the global attractor approach taken in this book for explaining and demonstrating cortical dynamics in memory and other cortical functions, such as sensorimotor control. As the mathematical object under consideration, namely, the neural firing rate model, is quite specific, we were able to avoid general, and rather abstract, concepts associated with the theory of noninvertible maps (e.g., Mira, 2007), and address the problem at hand in explicit terms. Furthermore, as shown in subsequent chapters of this book, reduction to circuits of identical neurons is justified by segregation of mutually asynchronous circuits into internally synchronous circuits either by the Hebbian paradigm or by synapse polarization. This reduction has, in turn, facilitated the direct application of a fundamental rule, namely, Descartes' rule of signs, which applies to scalar-valued polynomials. It might be noted in this context that a certain extension of Descartes' rule of signs to the multivariate case has been suggested (Bihan and Dickenstein, 2016), but does not appear to have a formal journal citation.

Part II

From Natural to Virtual Model Selection in Biological Motion Control

Chapter 8

Practical Realization of Brain Proximity Measures Toward Medical Treatment

8.1 Introduction

The previous chapters were essentially aimed at the mathematical nature of brain proximity measures. The basis and subsequent outcomes involved the selection of correct neural circuits for performing relevant activities. The present and following chapters are aimed at certain specific outcomes of such measures. Our main interest is directed at medical issues. Specifically, we address those difficulties associated with motion. The improvement of motion is facilitated by external means, which, in turn, activate neural circuits and select the most suitable neural circuit at a time. Hence, the focus of this and the following chapters is aimed at external, natural and mechanical systems, supporting the neurologically impaired.

In order to complement the linear dynamical models addressed in the subsequent chapters, this chapter shows that natural movement has mathematically nonlinear elements. Specifically, we separately address two different scenarios of natural movement. The first scenario is an imminent collision with a visual object, which is obviously unsafe in autonomous motion. Putting such an occurrence in the context of subsequent chapters, we represent visual objects by tiles and the event of collision by stumbling over a tile placed on the ground. Range, direction and time-to-contact can be extracted from the optical flow generated on the eye retina of a moving observer by the relative motion of rigid bodies (Koenderink and Doorn, 1975; Prazdny, 1981; Lee *et al.*, 1993). Such visual information may be used as a feedback signal to the movement control system for generating stumble-free movement (Baram, 1999). The second scenario addressed in this chapter is visual object targeting and approach. Addressed in a control-theoretic context, the issue was put to the author by a NASA colleague seeking an answer to the following question: How can an aircraft pilot, wishing to land on an airport runway, visually distinguish a real obstacle placed on the runway from a stain of oil? Recognizing the similarity of such a situation to that of natural bird preying resulted in a simple mathematical calculation, leading to a rather simple answer (Baram, 1996b).

8.2 Naturally Nonlinear Movement and Stumble Avoidance

Consider the observation geometry depicted in Fig. 8.1(a). The observer's eyes are first assumed to be placed at a constant distance from the floor, and the field of view is assumed to be centered about a line intersecting the floor at an angle θ. The size of the field of view is represented by the viewing angle α. The tile arrangement is shown in Fig. 8.1(b). The observer's motion in the "N" direction

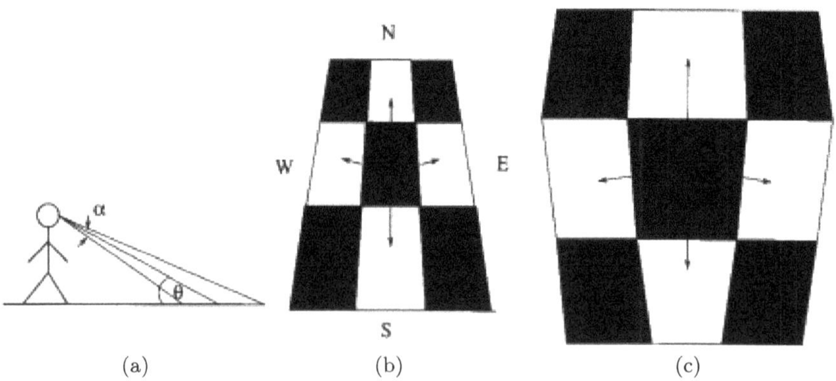

Figure 8.1. Tile observation geometry.

creates a motion, called "optical flow", of projected tile edges on the observer's eye retina. Each of the tile images expands. The motion of the projected edges of the center tile is represented by the arrows attached to the tile's edges. If the walker stumbles, actual flow will depend on the actual motion of the eye, but, in general, the tile's image expansion rate will increase drastically.

As will be shown mathematically, when the observer's velocity and eye level are constant, the net optical flow (NOF) of the center tile, representing the integral of the normal velocity along the projected tile boundary, grows approximately as the reciprocal of the forward velocity. As the observer moves forward, one center tile leaves the field of view and another enters. The walker fixates on the tile and visually tracks it, allowing the angle θ to grow with time. When θ reaches a certain threshold value, the observer's eyes make a fast movement to the next center tile, which lies farther away ("saccade", see, e.g., Hutton *et al.*, 1984), and the process repeats itself. The resulting NOF signal is a steady "spike train", as depicted in Fig. 8.2(a). It serves as a reference signal, which may be learned by the observer in a few steps of steady walk or from a standing position by eye motion, increasing the angle θ at a certain rate. A stumble will create an abrupt change in the actual spike train, as shown in Fig. 8.2(b), where the last spike is considerably greater than the previous ones. The difference is a single relatively large spike, as shown in Fig. 8.2(c). It constitutes a correction signal to the limbs, causing them to counteract a fall.

Clearly, even in healthy people, neither is the eye level constant, nor are the visual tracking of tiles and the catch-up saccades so smooth as to create a uniform spike train as the one displayed in Fig. 8.2. Yet, the proposed idealization seems to represent the essence of the stabilizing visual effect provided by the tiles. In Parkinson's patients, the idealization seems even less realistic (Hocherman and Aharon-Peretz, 1984; Klockgether *et al.*, 1995; Vasilakos and Beuter, 1993; Visser, 1983; Hutton *et al.*, 1984; Zaccara *et al.*, 1992), although certain eye-tracking functions, such as prediction in smooth pursuit, have been reported to remain intact (Fletcher and Aharpe, 1988). While the movement control impairment may cause severely irregular motion, the visual feedback signal, unsteady as it might be, provides, in essence, the same, and far more needed, stabilizing effect.

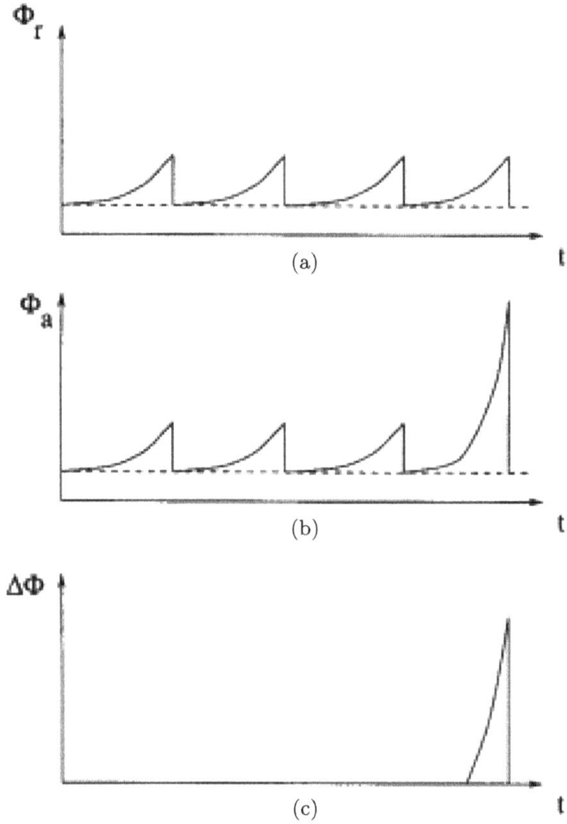

Figure 8.2. Spike trains of net optical flow generated by a steady walk (a) and by a stumble (b), and the difference signal (c).

It is like a cane which, held by a healthy walker, can provide some unnecessary comfort and, when held by a movement-impaired person, will provide, unsteady as it might be, essential support. The psychological effect is particularly interesting. It has been observed (Sacks, 1985) that the mere appearance of tiles in the field of view would cause a Parkinson's patient to start walking. This observation on the comforting effect of tiles seems to be supported by findings that more severely impaired patients display a higher anticipatory saccades activity than less severely impaired ones (Zaccara *et al.*, 1992). Relative tile motion generates such seemingly needed saccade activity. It has been suggested that pathological tremor present in Parkinson's patients interferes with their visual feedback

(Vasilakos and Beuter, 1993). Expanding and contracting visual tile images may be generated by tremor and may serve in creating a feedback signal that activates the muscles, which, in turn, counteract these disruptions.

A mathematical analysis of the visual information generated by motion with respect to a single tile and the processing of this information can be based on the obstacle detection mechanism (Ringach and Baram, 1994). The sequential appearance of different tiles in the field of view is handled by "saccades", as described above. Consequently, the corrective feedback signal is a sequence of "spikes", as depicted by Fig. 8.2(a), each representing the signal generated by a single tile. The tile projection geometry is shown in Fig. 8.3. A focal point O, representing the observer's location, is connected to a point P on a tile. An image plane, representing the retina, is placed at a certain distance from the focal point, perpendicularly to the line (OP), which intersects the image plane at point p. The points of the tile's boundary are connected to the focal point by straight lines, and the projected image of the tile is defined by the intersection of these lines with the image plane. Let (X, Y, Z) be a Cartesian coordinate system with origin at the focal point O, and let (x, y) be a coordinate system in the image plane, with origin at point p. Let the relative velocity of the observer, located at O, with respect to the tile be given by $V = (V_X, V_Y, V_Z)$. Normalizing the velocities, $(\tilde{V}_X, \tilde{V}_Y, \tilde{V}_Z) = (V_X, V_Y, V_Z)/D$, where D is the distance between

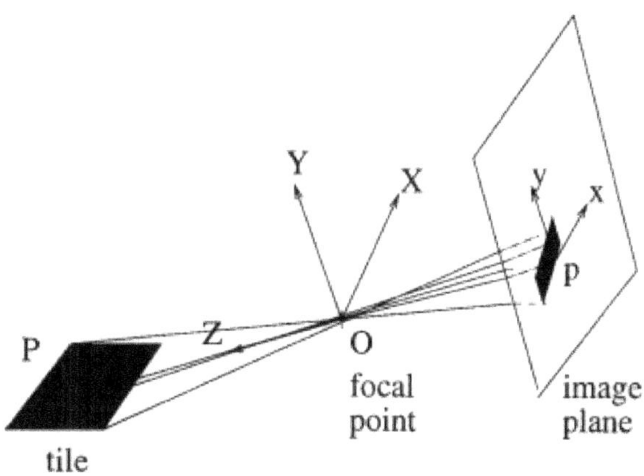

Figure 8.3. Tile projection geometry (Baram, 1999).

O and p, the optical divergence at a point p on the contour of the tile's projection on the image plane is given by (Ringach and Baram, 1994)

$$\Psi(p) \equiv \Delta \cdot v(p) = u_x + v_y = 2\tilde{V}_z + (\tilde{V}_x, \tilde{V}_y) \cdot (F_x, F_y) \tag{8.1}$$

where $v(p) = (u, v)$ is the velocity of p, u_x and v_y are the respective partial derivatives in the x and y directions, while F_x and F_y are the slopes of the tile with respect to the x and y directions. It can be seen that in a steady walk forward at a constant velocity V_z with $V_x = V_y = 0$, the local divergence is inversely proportional to the time to contact with the obstacle ($t = D/V_z = 1/\Psi(p)$). When the observer looks at a point on the floor well ahead and walks steadily, both V_z and ΔF are non-zero. A stumble will cause abrupt changes in V_z and F_y, increasing both.

For an untextured object, such as a tile, $\psi(p)$ may be integrated along the contour of the tile's image (Ringach and Baram, 1994). Let R denote the projection of the tile on the image plane, let $A(R)$ denote the area of the projected image, and let ∂R denote the projected boundary of the tile. Further, let ds and $d\ell$ denote infinitesimal elements of R and ∂R, respectively. Then the average divergence of the points of R is

$$\Phi(R) = \frac{1}{A(R)} \int_{R/\partial R} \Psi(p) ds = \frac{1}{A(R)} \int_{\partial R} v_n(p) d\ell = \frac{1}{A(R)} \frac{dA(R)}{dt} \tag{8.2}$$

where $v_n(p)$ is the velocity normal to the edge at p. We see that $\Phi(R)$ can be calculated from the component of the velocity normal to the boundary and that it represents the relative rate of expansion of the tile's image. The averaging action in Eq. (8.2) suggests that the noise in $\Psi(p)$ will be largely eliminated and that $\Phi(R)$ will be a more accurate measure of the reciprocal time to contact with the tile. As before, it should be noted that, in a steady walk, $\Phi(R)$ represents a certain rate of net optical flow through the tile's image used as a reference signal in the feedback control system. Between saccades, the observer, moving forward at a constant velocity, follows each of the center tiles with eyes. Consequently, the time to contact with the tile, as measured by the inverse of the average divergence,

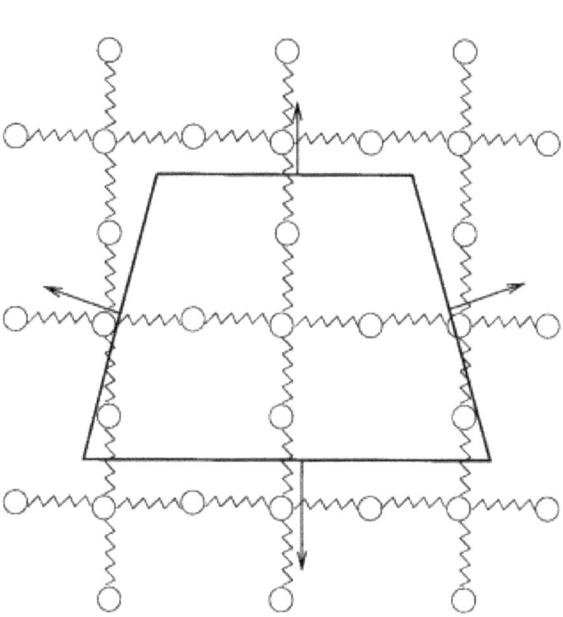

Figure 8.4. A locally connected neural network for calculating size change by diffusion.

reduces approximately at a linear rate equal to the forward velocity. This implies that the average divergence, or the net optical flow signal, increases between saccades at a rate which is inversely proportional to the forward velocity.

It has been shown (Ringach and Baram, 1994) that the calculation of $\Phi(R)$ is approximated by the final state of a network of locally connected array of cells, implementing a discrete diffusion process. Similar processes were previously proposed for performing other vision functions (Grossberg *et al.*, 1989; Grossberg, 1993; Grossberg and Todorovic, 1988). Assuming a two-dimensional grid arrangement of cells, as the one depicted in Fig. 8.4, the cell in position i at time $(k+1)\Delta t$, where Δt is some time interval, performs the calculation

$$u(i,\, k+1) = \frac{1}{|C(i)|} \sum_{p \in C(i)} u(p, k); \quad k = 0, \dots \tag{8.3}$$

where $C(i)$ is the group of neighboring cells connected to the ith cell and $|C(i)|$ is the number of cells in this group. This discrete diffusion process is initialized by zero everywhere but at the cells on the boundary of the image, which take the values (Marr and Hildreth, 1980)

$$v_n(i) = \frac{I_t(i)}{\sqrt{I_x^2(i) + I_y^2(i)}} \tag{8.4}$$

where I_t, I_x and I_y are the partial derivatives of the image light intensity with respect to t, x and y, respectively. It has been shown that, for a large number of nodes, the network is resistant to a high rate (nearly 50%) of connectivity failures (Ringach and Baram, 1992).

8.3 A Bird's-Eye View of the Descent Trajectory

When a bird descends toward a point on the ground for the purpose of landing or preying, it should be able to distinguish a three-dimensional object, constituting an obstacle or a target, from its two-dimensional background. The extraction of time to contact from optical expansion in birds was reported (Davies and Green, 1990; Lee *et al.*, 1993) and mechanisms for such calculations have been proposed (Nelson and Aloimonos, 1989; Ringach and Baram, 1994; Baram and Barniv, 1996). The simultaneous expansion of a background image can make target detection difficult. Following an earlier publication (Baram, 1996), we show next that the descent scenario facilitates a motion contrast between the optical images of a three-dimensional object and its two-dimensional background, making the distinction between them relatively easy.

Consider first a three-dimensional target at ground level and an observer, descending along a certain line, as shown in Fig. 8.5. The observer's image of the target is represented by the viewing angle μ, which increases monotonically, as long as the observer's altitude is greater than the height of the target. When the target is replaced by a flat background patch, as shown in Fig. 8.6, the viewing angle of the background patch, θ, will exhibit a different mode of behavior. Denoting the width of the background patch a, its horizontal distance from the observer x, and the observer's altitude h, θ may be written as the difference

$$\theta = \beta_1 - \beta_2 \tag{8.5}$$

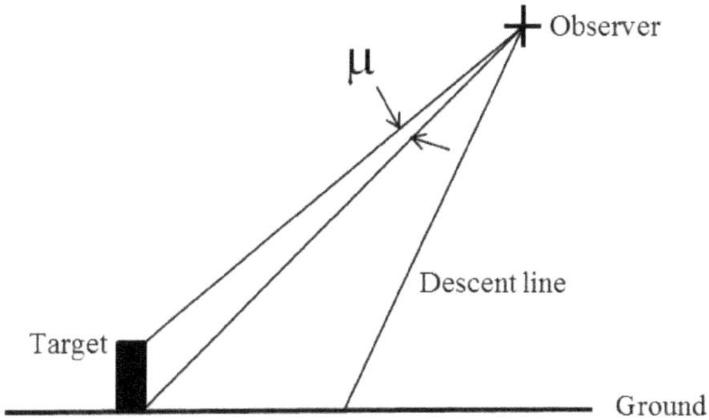

Figure 8.5. Target and observer geometry.

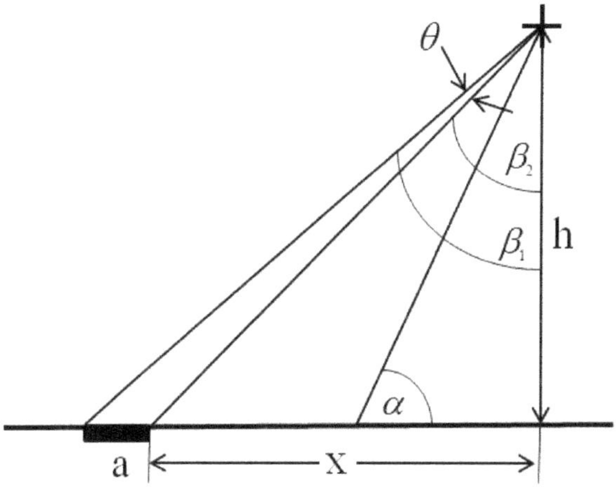

Figure 8.6. Flat background patch and observer geometry.

where β_1 and β_2 are the angles between the vertical to the ground at the observer's location and the straight lines connecting the observer to the far and the near ends of the patch, respectively. We have

$$\tan\beta_1 = \frac{x+a}{h} \quad \text{and} \quad \tan\beta_2 = \frac{x}{h} \tag{8.6}$$

hence

$$\tan\theta = \frac{ah}{h^2 + x^2 + ax} \tag{8.7}$$

Denoting, further, $b = x - h/\tan\alpha$, where α is the descent angle, we obtain

$$\frac{1}{\tan\theta} = \frac{1}{a}\left[\left(1 + \frac{1}{\tan^2\alpha}\right)h + (ab + b^2)\frac{1}{h} + \frac{a + 2b}{\tan\alpha}\right] \tag{8.8}$$

which relates θ to h along the line of descent, yielding a single maximum for θ at

$$h^* = \sqrt{\frac{b(a + b)}{1 + \tan^{-1}\alpha}} \tag{8.9}$$

The image of the background patch can be made to expand, contract, or remain stationary by choice of the descent angle α. Contraction or stationarity of the background's image will create a desirable contrast for the target's image, which keeps expanding in the vertical dimension. The greater the α is, the faster is the background contraction rate. However, in order to minimize the length of the descent trajectory, α should be as small as possible.

Selecting α locally, so that the size of h along the trajectory, as represented by Eq. (8.6), provides the maximum viewing angle θ of the background patch, will guarantee a maximal view of the background without losing the contrast between the target and its background. It might be further noted that, as in the case of obstacle detection discussed earlier, the image of the target is enhanced by an increase in the number of activated retinal and visual cortex cells, polarized positively by diffusion, as implied by image expansion, whereas the image of the background is progressively, albeit at minimal rate, silenced by negative polarization of the corresponding vision cells.

Taking then $h = h^*$ with $b = x - h/\tan\alpha$, and solving Eq. (8.4) for α, we obtain

$$\tan\alpha = \frac{h(1 + 2x)}{x^2 + ax - h^2} \tag{8.10}$$

The descent trajectory may be calculated for a background patch of a certain standard size, say $a = 1$ (equivalently, all lengths may be assumed, without loss of generality, to be given in units of a).

It will then take the form

$$\frac{dh}{dx} = \frac{h(1+2x)}{x^2 + x - h^2} \tag{8.11}$$

which together with the initial condition(x_0, h_0) defines the descent trajectory.

Our analysis was based on the assumption that the local descent angle α is greater than $\tan^{-1}(h/x)$. This is clearly a necessary condition if the observer is to meet the target, with α changing monotonically. It follows from Eq. (8.11) that, for the latter condition to hold at the initial point of the trajectory, we must have $h_0 \leq x_0$. If this is not the case, the ratio x/h must be increased to 1 first. A bird may simply drop straight down to the point $h = x$, letting gravitation alone take effect. Then, spreading wings, it may follow the trajectory represented by Eq. (8.11).

Trajectories are shown in Fig. 8.7 for $h_0 = 50$ with $x_0 = 50$ (red), $x_0 = 100$ (blue) and $x_0 = 200$ (green). In the first case, the initial descent is implemented by a drop straight down. It can be seen that in all cases a nice descent trajectory, ending at the target location, is attained.

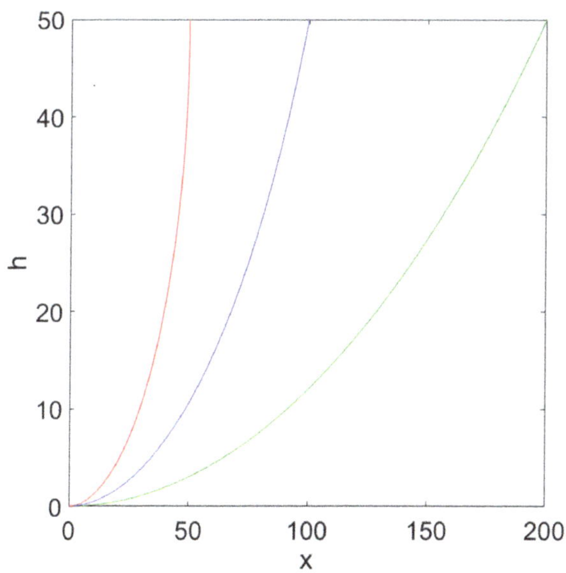

Figure 8.7. Descent trajectories for different initial conditions.

It might be noted that the trajectory that maintains a constant viewing angle of the background patch in a global sense is the relevant part of the circle which passes through the initial location of the observer and the two ends of the patch. There are, however, two major problems associated with flying along such a circular trajectory: It cannot be calculated locally from optical measurements alone, and it intersects the ground at a non-zero angle (which might have a disagreeable effect on any bird). In contrast, the proposed trajectory can be derived locally by simply varying the descent angle until the viewing angle appears stationary. In terms of Fig. 8.7, the boundary of the image is replaced by that of the background patch viewed by the bird, and the diffusion process, causing changes in cell polarity, stops. Such local stationarity allows a gradual decay of the viewing angle. More importantly, the descent angle safely nullifies at zero altitude.

8.4 Discussion

Movement execution is perhaps the most fundamental task of the brain. Movement has been associated with enhanced neurogenesis, as well as learning and memory (van Praag *et al.*, 1999). Examining two specific natural movement tasks, one aimed at obstacle avoidance and the other aimed at object targeting, we have seen that, in contrast to the linear nature of the technologically driven control processes addressed in other parts of this book, the natural processes addressed in this chapter are largely nonlinear. The nonlinear effects do not reduce the accuracy of the resulting movement. Quite the contrary, they generate the desired outcome with mathematical precision. On the other hand, neurological impairments, as addressed in the following sections, call for technological visual and auditory support, which is largely linear.

Chapter 9

Virtually Linear Biological Motion Control

9.1 Introduction

The previous chapter has addressed certain cortical processes which propagate sensory information so as to achieve natural movement objectives, specifically, object avoidance, targeting and approach. The underlying mathematics of the control mechanisms was nonlinear. We now show that the inclusion of a technological mechanism, specifically, a portable virtual reality device, produces, by linear feedback control, improved gait entrainment in the neurologically impaired. Notably, model selection for improved movement, addressed mathematically in Chapters 9–14, is shown to be provided by cortical operations, specifically, vision, hearing and motion control. We first mention a certain historical development, which, based on a scientifically sound early proposition, namely, visually-based feedback control, has later evolved in a fundamentally false direction, namely, open-loop control, before finding its way to highly effective, technologically sound feedback control.

A seminal study (Martin, 1967) and subsequent clinical tests (Bagley *et al.*, 1991; Morris *et al.*, 1994a, 1994b; Azulay *et al.*, 1999) have shown significant improvement in gait parameters of patients with Parkinson's disease (PD) walking over transverse lines drawn on the ground. Early attempts to generate such visual cues by portable virtual means have resulted in open-loop systems, imposing, by a constantly moving visual cue (Prothero, 1993; Weghorst *et al.*, 1994;

Riess and Weghorst, 1995) or a geometrically patterned treadmill (Hanakawa *et al.*, 1999), a constant walking speed on the patient. A comparison of open-loop visual cuing by virtual means to transverse line markings on the ground (Griffin *et al.*, 2011) has found the first to have a marginal effect and the second to have a significant positive effect on gait parameters in PD patients. External open-loop auditory entrainment of gait, where a rhythmic sound is imposed on the patient in a metronome-like fashion, has been studied (Kritikos *et al.*, 1995; Thaut *et al.*, 1996; McIntosh *et al.*, 1997; Howe *et al.*, 2003; Nieuwboer *et al.*, 2007). Yet, a need for constant vigilance and attention strategies to prevent reversion to impaired gait patterns caused by repetitive stimuli has been noted (Morris *et al.*, 1996). It has been further suggested that external rhythmic entrainment instigates dopamine reward (Salimpoor *et al.*, 2015), seemingly supporting open-loop intervention. However, open-loop control is known to be inherently unstable, with potentially disastrous consequences due to error accumulation (Kuo, 1981).

In sharp contrast, closed-loop feedback systems, when correctly designed, can regulate and stabilize otherwise unstable dynamics (Kuo, 1981). Analysis has shown that walking over earth-stationary visual markings constitutes a closed-loop feedback control system that stabilizes and regulates gait (Baram, 1999). Portable virtual realization of autonomous gait entrainment by sensory feedback, facilitated by the gradual miniaturization of sensing, computation and display technologies, has led to the development of a closed-loop sensory feedback device for gait improvement in the neurologically impaired (Baram, 2004). While the effect created by the device is similar to that created by walking on a real floor with visual markings, the virtual earth-stationary visual markings are location-independent, as the patient is, in principle, free to go anywhere, making such entrainment completely autonomous.

9.2 Linear Control and Autonomous Cortical Model Selection

An examination of the natural sensory-motor control system underlying human locomotion with respect to the visual scenery reveals that it is the physical motion of the body that generates the visual cue and not the other way around (Baram, 1999). This observation is crucial to understanding the difference between external entrainment and autonomous entrainment of gait. The first represents a control system operating in open loop, while the second represents a closed-loop feedback control system. The two control paradigms are illustrated in Fig. 9.1. In external, open-loop entrainment, a visual cue, moving at a constant speed, and/or a rhythmic auditory cue are generated artificially and fed through the eyes and/or the ears to the cortical motor lobe, which activates the limbs so as to respond to the sensory cues. On the other hand, in the autonomous closed-loop entrainment, an artificial realization of the natural sensory-motor control system is a closed-loop feedback system, where the generation of the sensory feedback cue is controlled and regulated by sensory and movement control cortical elements. Here, the resulting motion of the visual cue and the rhythm of the auditory cue are matched to the motion of the body, which is, normally, matched to the steps executed by the legs. When there is no motion of the body, there is no sensory cue. The linear nature of both systems, open or closed loop, is characterized by the products of matrices, which may, for simplicity, be materialize in scalar forms. The overall effects are products of a linear operator.

The safety of autonomous gait entrainment is ratified by control-theoretic considerations: closed-loop feedback systems, correctly designed, are, due to error correction, inherently stable, while open-loop ones, externally driven, systems are, due to error accumulation, inherently unstable and, hence, unsafe (Kuo, 1981). As can be seen from Fig. 9.1, while error correction and noise elimination are made possible by the sensory feedback path of autonomous entrainment, these cannot be implemented by the open-loop (no feedback) system of external entrainment.

The open-loop gait control system has the linear formula

$$x_n = Au_n + \omega_n$$

where

$$A = A_1 A_2 A_3 A_4$$

with $A_1, \ldots, A_4$ linear operators representing the corresponding components. While the first two, rectangular-shaped ("image/sound generator" and "display/phones") components are mechanically

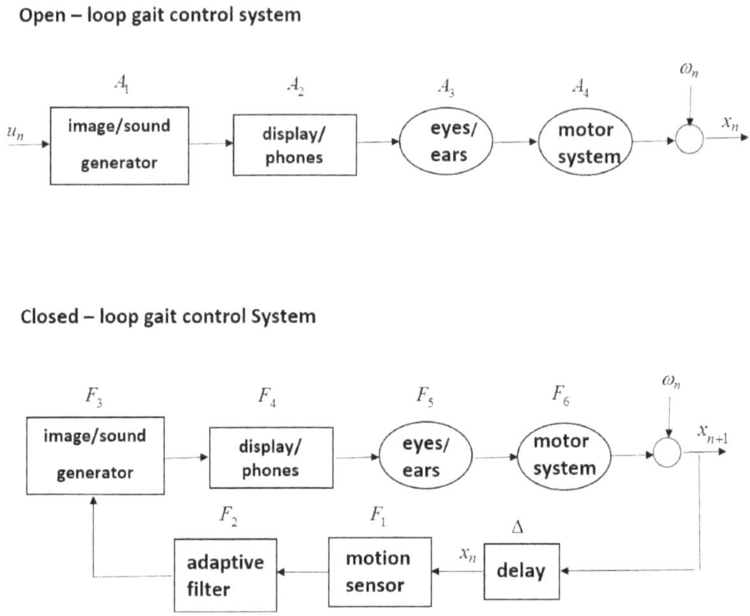

Figure 9.1. Autonomous (closed-loop) vs. external (open-loop) entrainment of gait.

constructed, the last two, oval-shaped ("eyes/ears" and "motor system") components are cortical.

It should be noted that a neurologically impaired person is subject to a variety of both internal and external circumstances. These may be ruled by personal states of health as well as changes in motion targets (hence, corresponding brain proximity measures).

The closed-loop gait control system has the linear formula

$$x_{n+1} = Fx_n + \omega_n$$

where

$$F = \Delta F_1 F_2 F_3 F_4 F_5 F_6$$

with Δ the delay operator yielding

$$x_n = \Delta x_{n+1}$$

and $F_1, \ldots, F_6$ linear operators representing the corresponding components. Here, while the first five components are mechanically constructed, the last two oval-shaped ("eyes/ears" and "motor system") components are, again, cortical.

While the open-loop system is driven by a constant external input, the closed-loop system is driven by the patient's own forward motion. At the same time, while the effect of the external noise, ω_n, on the outcome cannot be corrected by the open-loop system, it is corrected by the feedback in the closed-loop system.

The autonomous model selection is controlled by the natural adjustment of the biological cortical components, namely, the "eyes/ears" and the "motor control" components. The biological construct of the two cortically controlled components has far-reaching capabilities of adapting to environmental, biological and medical changes. Such adaptation corresponds to the notion of model selection, addressed mathematically in Chapters 8–12. The remaining components of the system are technologically contrived and constructed and, for the present considerations, are assumed to be unchanged during movement. Each box, albeit highly complex, may be approximately represented by a reduced linear system.

9.3 Virtual-Natural Composition of the Motion-Control Aid for Neurologically Impaired Gait

As depicted in Fig. 9.2, a portable sensory feedback device, employing inertial sensors, an adaptive filter and a microprocessor contained in a belt-mounted cellphone-size box, is connected to a see-through micro-display, generating an earth-stationary visual feedback cue in response to gait entrainment (Baram, 2004). In addition to the visual feedback cue of glide-symmetric virtual tiles delivered by the display in coordination with the user's steps, the device also produces an auditory feedback cue in the form of a clicking sound delivered through earphones in response to every step taken by the patient. In contrast to open-loop, metronome-like devices, which attempt to impose a walking pace on the patient by a constant auditory cue, the feedback device produces an auditory cue matched to the walking pattern. A balanced, steady walk generates a rhythmic auditory cue. Any deviation from such a gait pattern will result in a deviation from the auditory rhythm and will be corrected by a change of the gait pattern in a feedback fashion. The head-mounted display and earphones bring the sensory feedback signals closer to the sensors, the eyes and the ears, making the sensory effect more pronounced

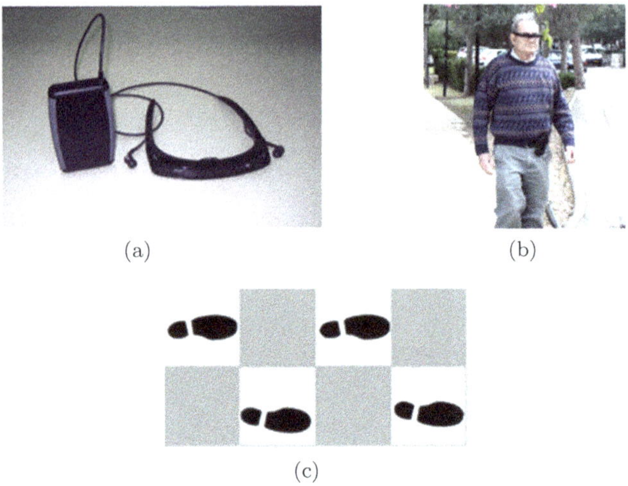

Figure 9.2. (a) Autonomous entrainment device, (b) walking with the device, and (c) stepping on micro-displayed evenly spaced glide-symmetric virtual tiles.

and easier to follow. An open-loop capability, producing constant movement of the visual cue and a constant rhythmic auditory cue, was also added to an early version of the device for experimental comparison purposes. In contrast to the naturally nonlinear control mechanisms mathematically and graphically presented in the previous chapter, the device virtually generates both visual and auditory linear feedback control. The linearity of the visual mode is manifested by step coordination with evenly spaced glide-symmetric virtual tiles, while the linearity of the auditory mode is manifested by rhythmic clicking sound generated by the contact of each of the steps with the corresponding virtual tiles.

9.4 Dopamine Reward and Reward-Seeking

Before embarking on the presentation of clinical study results in subsequent chapters, it seems appropriate to pay attention to a certain internal cortical agent, which is believed to have a fundamental role in the well-being and physical performance of both healthy individuals and neurologically impaired patients. Such issues are believed to be associated with the presence, or lack thereof, of a neurotransmitter called dopamine. A highly insightful study (Jenkinson and Brown, 2011) suggests that the frequency level of beta-range neuronal activity in the basal ganglia is modulated by the level of dopamine discharge at sites of cortical input, which are, in turn, modulated by salient internal and external cues. Moreover, it suggests that the level of the beta-range frequency provides a measure of the likelihood that a new voluntary action will be actuated. Put in our context, this insight suggests that autonomous entrainment of gait is more highly rewarded by dopamine discharge than rhythmic external cues. Yet, dopamine depletion in basal ganglia neurons of patients with PD suggests that it may be reward-seeking, rather than actual dopamine reward, which comes into play in gait entrainment of patients with PD.

It has been further suggested that external open-loop entrainment instigates dopamine reward (Miendlarzewska and Trost, 2014; Salimpoor *et al.*, 2015), seemingly supporting open-loop intervention. However, as implied by clinical results, external entrainment, having some positive effects on some of the patients, also results in adverse effects, such as freezing of gait in patients with PD. These mixed results may be characterized as the "double-edged sword" of external entrainment.

It has been hypothesized that dopamine neurotransmission, highly evidenced in limbic areas of the basal ganglia (DiChiara *et al.*, 1994), elicits exhilaration or excitement in response to novel stimuli (Cloninger, 1987). The dopamine receptor gene D4DR has been linked to novelty-seeking behavior (Ebstein *et al.*, 1996; Benjamin *et al.*, 1996). In the context of cognition, novelty has been associated with "the ability to think and act independently" (The Free Dictionary, 2018). While it has been suggested that novelty-seeking, representing intellectual curiosity, aesthetic sensitivity and risk taking, is low in dopamine-deficient patients with PD (Menza *et al.*, 1993), highly pronounced novelty-seeking behavior has been found in patients with PD, regardless of medication status (Djamshidian *et al.*, 2011). Dopamine reward has also been associated with vigilance (Aston-Jones *et al.*, 1994; Ikemoto, 2007), defined as "alert watchfulness" (The Free Dictionary, 2018).

While movement has been widely associated with reward, specifically, dopamine discharge by basal ganglia neurons, movement improvement in PD patients suggests the existence of a bypass mechanism. Prior reward learning, possibly facilitated by earlier normal preconditioning or medication, provides response incentive without actual dopamine signaling (Wassum *et al.*, 2011). In other words, learning to select responses that lead to reward results in reward-seeking behavior (Graef *et al.*, 2010; Moustafa, 2010). As novelty and vigilance have been widely recognized as key proponents of dopamine reward, the level of dopamine reward or reward-seeking should correspond to the levels of novelty and vigilance. While neither novelty nor vigilance has generally accepted measurable manifestations, both seem to be intuitively well-understood notions. In a cognitive perspective, novelty has been characterized as the ability to think and act independently (The Free Dictionary, 2018). In this respect, novelty does not seem to apply to external entrainment, which enforces a rhythmic auditory cue, or a constantly moving visual cue, on the patient. Indeed, it has been noted that the failure of external entrainment to produce satisfactory gait control in most patients is caused by the perpetual motion of the image, which, being unaffected by the patient, is eventually neglected (Morris *et al.*, 1996). In sharp contrast, autonomous entrainment, allowing the patient to decide if and when to take a step and at what stride length and speed of walk, is an embodiment of independent, free-willed motion. Yet, vigilance appears to play a role in both external and autonomous entrainments, as both motivate a certain correspondence between step preparation and execution, on the one hand, and the presentation of a sound or an image, on the other. The unequal balance between novelty and vigilance in the two cases suggests that while both external and autonomous entrainments present cases for dopamine reward, or reward-seeking, autonomous entrainment, instigating both novelty and vigilance, may present a much stronger case than external entrainment, which instigates vigilance only.

9.5 Discussion

The findings reviewed and compared in this chapter unambiguously show that autonomous entrainment of gait by visual and/or auditory feedback is not only decisively safer but also significantly more effective than external entrainment. Specifically, it is the difference in retinal sensor polarity, generated by visual images, which entrains the motor system in a feedback fashion. While external gait entrainment is a "double-edged sword", producing adverse effects such as freezing in Parkinson's patients, autonomous entrainment comes without such adverse effects. While external entrainment represents an open-loop control system which is, due to error accumulation, inherently unstable and, hence, unsafe for patient use, autonomous entrainment constitutes a closed-loop feedback control system which is, by error correction, inherently stable and, hence, safe for patient use. Autonomous entrainment of gait by visual or auditory feedback produces significantly higher improvement in gait parameters than external entrainment. The checkerboard tiles geometry, being two-dimensional, produces in autonomous entrainment significantly higher levels of novelty and vigilance, resulting in significantly higher improvement in gait parameters than the one-dimensional transverse lines geometry. Yet, the level of gait improvement can vary significantly, not only with respect to the nature of the disorder but also with respect to disorder severity and the patient's own biological and biographical attributes, such as age, education, mental state and cognitive abilities. Finally, it might be noted that hundreds of publications related to gait entrainment, while referring to autonomous (or closed-loop) entrainment, have, in fact, addressed external entrainment without making a distinction between external and autonomous entrainment. We find such presentations to be highly, although not necessarily intentionally, misleading. Following an early study presented in the following chapter, where we compared the gait improvement in Parkinson's patients employing the open and closed-loop versions of the device described above, all subsequent studies reported in the remainder of the book address closed-loop control alone.

Part III

Gait Entrainment by Sensory Feedback in the Neurologically Impaired

Chapter 10

Walking on Virtual Tiles with Parkinson's Disease

10.1 Introduction

Gait impairment is one of the main complaints of advanced Parkinson's Disease (PD) patients. Such patients often walk slowly with shuffling and dragging steps, reduced stride length, stooped posture and diminished arm swing. Certain sensory stimuli are known to improve the walking abilities in PD patients to a considerable extent. A pioneering study (Martin, 1967) has reported positive effects of earth-stationary visual cues, such as transverse lines, on gait in PD patients. While the positive effect of a walking stick as an aid to locomotion in patients with PD has been noted (Dunne *et al.*, 1987), the utility of visual markers has been further described by several investigators (Bagley *et al.*, 1991; Morris *et al.*, 1994, 1996; Azulay *et al.*, 1999). However, visual markers cannot be placed along each path the patient takes. Beneficial effects on gait in two PD patients using visual cues displayed by a portable head-mounted device have been reported (Prothero, 1993). The images were either stationary or in constant perpetual motion, unrelated to the patient's own dynamics (the patient sees the same monotonous cue whether standing or walking, looking up or down). Clearly, this open-loop approach cannot provide the safety needed by PD patients for movement. In contrast, the closed-loop effects provided by real-world visual cues on the regulation and stabilization of gait (Baram, 1999) provide the necessary safety. Here, we first present the results of the ground-breaking work

(Baram *et al.*, 2002), showing that such effects must be included in a display of virtual cues in order to achieve a significant improvement in the walking abilities of PD patients. We then present the results of subsequent studies, ratifying our original observations.

10.2 Seminal Study of Gait Entrainment in PD by Virtual Tiles

The original device used in this study comprised a head-mounted three-axis rotational accelerometer, a body-mounted three-axis translational accelerometer and a see-through head-mounted display, all connected to a wearable computer. It could be operated in two modes: the open-loop mode, in which the virtual tiled floor performs perpetual motion toward the observer at a constant speed, unrelated to the patient's motion or lack thereof, and the adaptive closed-loop mode, in which the virtual floor appears to be fixed in space, as a real floor. In both cases, the virtual floor is superimposed on the real world, seen through the display. In contrast to the open-loop system, depicted in the upper part of Fig. 10.1, the closed-loop system, depicted in the lower part of Fig. 10.1, modifies the image of the virtual tiled floor produced by the image generator, in accordance with the patient's motions, measured by a body-mounted forward accelerometer.

Fourteen PD patients of mean age 68.2 ± 8.17 participated in the study. Mean disease duration was 9.36 ± 6.64 and mean clinical severity according to HY staging (Hoehn and Yahr, 1967) was 3.04 0.84. All patients had visual acuity 6/6 with correction when necessary. All tests were performed at the Movement Disorders Clinic, Rambam Medical Center, Haifa, Israel, around the same time in the morning, following 12 hours off medications. Before starting a test, each patient was made familiar with the device. Each test consisted of the patient walking a straight track of 10 meters four times. At the start of each test, the patient was verbally instructed to start walking. The time to complete the track and the number of steps for each path were recorded. In the first test (the reference test), the display was turned off. In the second test, the open-loop system was turned on, displaying a virtual tiled floor in perpetual motion toward the observer at the maximal speed level, which was still comfortable for the patient. The third test employed the adaptive closed-loop system. The order of the second and third tests was then reversed, and the results for each of the tests were averaged in order to eliminate the effect of training from the comparison. The patients wore the device even when the display was turned off so as to eliminate the burdening effect of the device from the comparison.

The test results are given in Table 10.1, where the last four columns show the percentage changes in the performance parameters obtained for the open-loop and the closed-loop systems with respect to the reference test. It can be seen that, in all patients but one, performance was significantly improved (higher speed, longer stride) with respect to the reference test when the closed-loop system was turned on. This measured improvement also represents the observed improvement in the quality of the steps. Patients who dragged their feet on the ground in the reference test raised them noticeably higher when the closed-loop system was turned on. The one patient who was not helped by the display at all, MR, had no walking impairment (as can be seen from his test parameters, he had, by far, the best performance of all patients in the reference test). The last two rows in the table show the average performance of our patient group (excluding MR, who, as noted before, had no walking impairment (as can be seen from his test parameters, he had, by far, the best performance of all patients in the reference test). It can be seen that, on average, the proposed closed-loop system improves performance.

It can be seen that, on average, the proposed closed-loop system improves performance by about 26% (walking speed) or 31% (stride length) with respect to the reference test. It should also be noted that the standard deviations of these results are rather high, which implies that the results should be evaluated mainly on an individual basis. Certain Parkinson patients would be helped by the proposed approach to a very significant degree (50–100%), while others would be helped to a lesser degree. Few, in particular those with no walking impairment, would not be helped at all. The average results for the open-loop system are about half as good as those for the closed-loop system. Furthermore, the standard deviations for the open-loop system with respect to the means are drastically higher than those for the closed-loop system and the reference state. This means that the open-loop system affects different patients in a drastically different way (indeed, for some of the patients, the open-loop system gave

Table 10.1. Performance results of PD patients.

				Display off		Open-loop		Closed-loop		Open-loop (Improvement%)		Closed-loop (Improvement%)	
Patient	Age	Yd	HY	Walking speed (m/s)	Stride length (m)	Walking speed (m/s)	Stride length (m)	Walking speed (m/s)	Stride length (m)	Walking speed (m/s)	Stride length (m)	Walking speed (m/s)	Stride length (m)
1	74	4	2	0.881	0.465	1.063	0.541	1.107	0.571	20.2	16.3	25.2	22.9
2	72	2	2	0.837	0.339	0.738	0.370	0.772	0.385	15.9	9.1	21.2	13.5
3	77	1	2	0.845	0.435	1.056	0.556	1.272	0.667	25.0	27.8	50.5	53.3
4	67	20	3	0.416	0.256	0.575	0.111	0.535	0.392	38.2	62.9	28.5	52.9
5	61	2	2.5	0.784	0.541	0.918	0.400	0.855	0.588	17.1	2.8	9.1	8.8
6	66	15	4	0.314	0.168	0.073	0.488	0.339	0.220	−77	−34	8.0	30.7
7	61	15	3	0.686	0.323	0.887	0.444	0.820	0.408	29.3	24.0	19.5	26.5
8	76	5	4	0.774	0.388	0.144	0.500	0.907	0.426	47.8	25.6	17.2	9.6
9	73	11	3	0.695	0.408	0.881	0.476	0.791	0.417	26.8	8.9	13.8	−2.1
10	57	3	2	0.620	0.385	0.907	0.400	0.887	0.488	46.2	30.0	43.0	26.8
11	73	10	3	1.038	0.555	0.855	0.488	0.888	0.500	−18	−14	−14	−9.9
12	82	20	4	0.508	0.339	0.601	0.140	0.590	0.385	18.3	18.0	16.0	13.5
13	59	10	4	0.773	0.435	0.864	0.434	0.895	0.488	11.8	12.2	15.9	12.2
14	57	13	4	4.405	0.154	0.243	0.145	0.668	0.531	−40	−9	65.5	12.8

Note: yd: years of disease duration; HY: Hohen−Yahr staging.

slightly better results than the closed-loop one). The behaviors of JS and NM are particularly noteworthy in this respect. Both these patients were severely Bradykinetic, unable to stand or start walking on their own. When the closed-loop display was turned on, both these patients were able to initiate and sustain a near-normal unaided walk. When the open-loop system (perpetual motion display) was turned on, both patients were able to initiate a walk, but then experienced freezing episodes. It can be seen that, for both patients, the performance parameters for the open-loop system are drastically lower than those for the closed-loop system and even those for the reference test. Some patients reported discomfort, dizziness and nausea caused by the perpetual floor motion of the open-loop system. Most patients reported relative comfort with the self-activated, closed-loop adaptive system and indicated a clear preference for it over the open-loop system.

10.3 Recorded Cortical Sensory Motor Activity

An early study (Hanakawa, 1999), employing blood-flow measurements, has examined cortical activity in patients with PD assisted by treadmill-induced open-loop visual cues for gait entrainment. It has been found that the right lateral pre-motor cortex, which is mainly regulated by cerebellar inputs, was activated to a greater extent in PD patients than in age-matched healthy individuals. On the other hand, the healthy individuals activated mainly the supplementary motor area (SMA), which was underactivated in PD patients.

A later study (Velu *et al.*, 2013), employing electroencephalography (EEG) recordings, has examined the cortical activity in patients with PD assisted by closed-loop visual feedback induced by a portable augmented-reality device, as described in the previous section. This case study has found a considerably higher increase in the beta frequency range of the connections between the occipital and the motor regions and between the occipital and the parietal regions of a responding patient with PD, compared to a non-responding patient and to an age-matched healthy individual. Revisiting the results of that study, we show here that they concur with the observations on cortical polarity and singularity segregation reported in this book.

This is particularly noteworthy in view of the fact that polarity and singularity segregation have been argued in this book on predominantly mathematical grounds. The cortical recordings serve as direct and vivid evidence of not only the validity of cortical polarity and singularity segregation but also their specific functional role in the directional connectivity between brain areas associated with the cortical functions of interest, specifically, vision, navigation and movement.

Electroencephalography (EEG) is an electrophysiological monitoring method to record the electrical activity of the brain. Employing a portable EEG device, cortical activity, relating neural firing rate to scalp-localized potential values, was recorded and compared in a cortical network composed of occipital (Oz), parietal (P4) and motor (Cz) channels (Velu *et al.*, 2013). The choice of these channels was based on anatomical findings from prior fMRI and PET experiments on paradoxical gait in PD (Hanakawa *et al.*, 1999). A patient with PD responding to visual autonomous entrainment (PDr), a non-responding patient with PD (PDnr), and a healthy age-matched individual (Control) participated.

Renormalized partial directed coherence (RPDC), representing directional connectivity between the cortical regions of interest, was used, as it provides a scale-free estimator, avoids normalization by outflows, and is not restrained by frequency-dependent statistical significance (Schelter *et al.*, 2009). The experiment consisted of three walking stages: A (visual feedback turned off), B (visual feedback turned on), and C (visual feedback turned off).

The EEG spectra show color-coded RPDC differences between stages C and A in the delta (0–4 Hz), alpha (8–12 Hz), and beta (12–30 Hz) frequency bands corresponding to the connection between Oz and Cz (Fig. 10.1) and the connection between Oz and P4 (Fig. 10.2). Blue indicates a decrease, while red indicates an increase in the differential RPDC (RPDC(C)-RPDC(A)). It can be seen that there is increased RPDC between Oz and Cz (Fig. 10.1) and between Oz and P4 (Fig. 10.2) in the beta range in PDr, reduced in PDnr and only weakly noticeable in Control. Similar results are seen in the difference in activity between stages B and A, but we only show the differences in RPDC between stages C and A to avoid any confound from neural responses to the presence of visual cues.

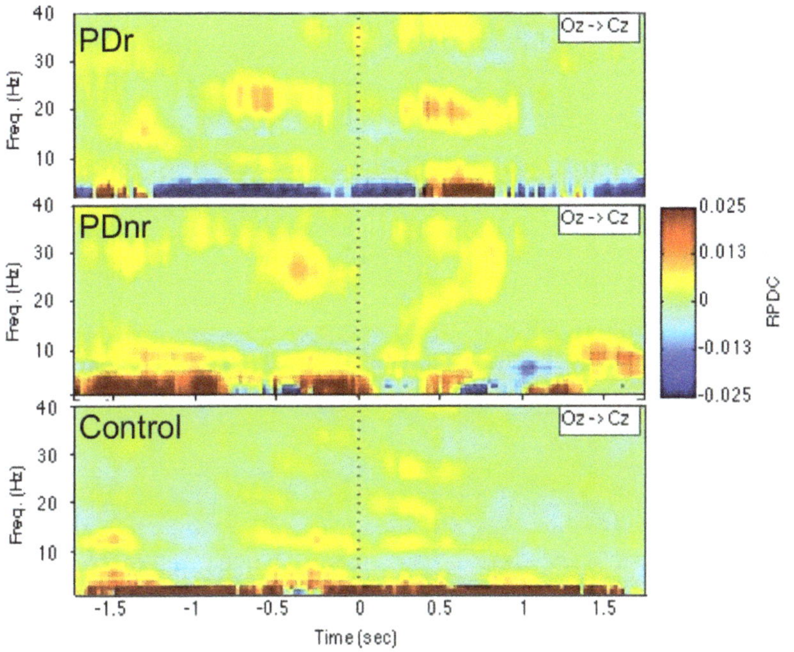

Figure 10.1. PD patient's color-coded RPDC, representing direction-specific connectivity from occipital region (Oz) to motor region (Cz). Blue indicates a decrease in RPDC, while red indicates an increase in RPDC between stage C and stage A. There is an increased RPDC in the beta frequency range in PDr, reduced in PDnr, and only weakly noticeable in Control (Velu *et al.*, 2013).

Putting the EEG results in the context of cortical polarity, red-colored RPDC represents positive polarity in the connectivity between the respective regions (Oz to Cz and Oz to P4), while blue-colored RPDC represents negative polarity. It can be seen that in Fig. 10.1, showing Oz to Cz RPDC, there are positive polarity (red-colored) signatures in the beta frequency range (12–30 Hz) corresponding to PDr, considerably reduced in the beta range corresponding to PDnr, and almost non-existent in the beta range corresponding to Control. On the other hand, there is predominantly negative polarity in the delta frequency range corresponding to PDr, predominantly positive polarity in the delta range corresponding to PDnr, and considerably reduced in the delta range corresponding to Control. There is a very weakly noticeable activity in the alpha

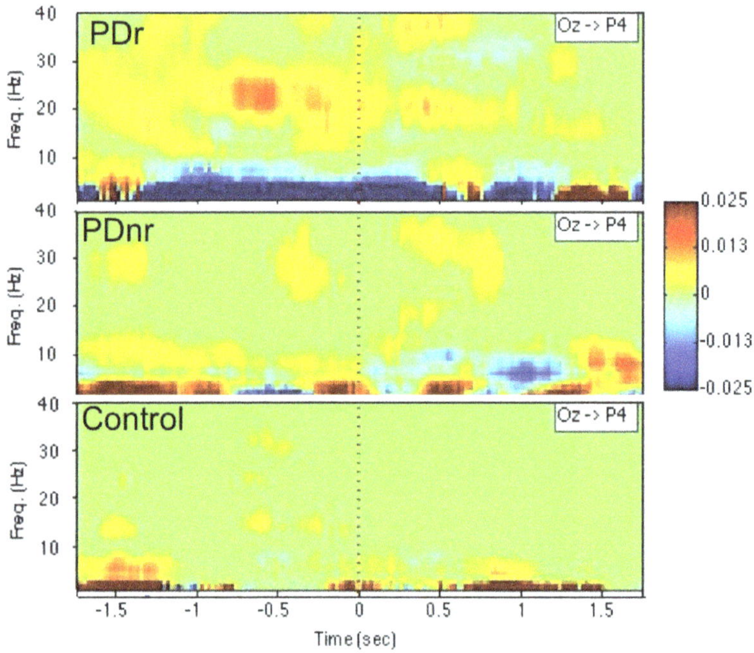

Figure 10.2. PD color-coded RPDC, representing direction-specific connectivity from occipital region (Oz) to parietal region (Cz). Blue indicates a decrease in RPDC, while red indicates an increase in RPDC between stage C and stage A. Occipital region (Oz) to motor region (Cz) flow shows increased RPDC in the beta frequency range in PDr, reduced in PDnr, and only weakly noticeable in Control (Velu *et al.*, 2013).

frequency range corresponding to all three participants. Essentially similar results can be seen in Fig. 10.2, showing Oz to P4 RPDC. It might be noted that in the beta range for PDr, there are positive (red) RPDC signatures around both 0.5 s and −0.5 s in both Figs. 10.1 and 10.2. These signatures appear to correspond to two consecutive steps, about 1 s apart within the recorded walking sequence, centered at 0 s.

In the context of segregated singularity, the visible differences between the RPDC signatures corresponding to the different frequency zones, delta, alpha and beta, representing different connectivity sequences, indicate segregation between the modes of connectivity and, hence, segregation of the corresponding cortical circuits. As shown in this book, a difference in circuit connectivity would imply a difference in circuit firing dynamics, as, indeed, represented by the EEG activities corresponding to the three frequency domains. In closing, we note, again, that it was not our purpose here to point out the advantages of visual feedback for improvement of PD gait, which was done in the original study (Velu *et al.*, 2013), but, rather, to point out the presence of cortical polarization and singularity segregation, mathematically revealed and analyzed in this book and ratified by the EEG recordings.

Case studies are a highly valuable means for ground-breaking research. Their merits include the following: detecting novelties, generating hypotheses, applicability when other research designs are not implementable, allowing emphasis on the narrative aspect (in-depth understanding), and educational values. However, following are the major limitations: lack of ability to generalize or to establish the cause-effect relationship, danger of overinterpretation, retrospective design, and distraction by focusing on the unusual. Their results, while understandably subject-dependent, can be rather conclusive. The results presented and analyzed in this chapter represent a positive case in point. Each of the subjects tested represents a widely recognized category. Among patients with PD, there are patients who respond to treatment and patients who do not. Controls are, normally, age-matched healthy individuals who do not show a noticeable response to treatment. As the purpose of the study discussed here was to experimentally examine a theoretically founded hypothesis, regarding, essentially, a model, and, only remotely, a cure for a disease (which has been examined in clinical studies of large cohorts of patients before), a case study was an important first step. The conclusive results were highly ratifying of the theory tested here: cortical activity is driven by circuit polarization, and its dynamic singularities are highly segregated. Motivated readers and researchers are highly encouraged to conduct and expand such studies.

10.4 Follow-Up Clinical Test Results

Following the seminal study (Baram *et al.*, 2002), we have developed a more advanced portable miniature virtual reality device and tested it on a group of PD patients at the Carmel Medical Center in Haifa, Israel (Badarny *et al.*, 2014). The essential parameters examined were the same as those examined in the seminal study, with the additional testing of the residual effect (without the device) 15 minutes after the device-guided tests. The details of this follow-up study are described in the following.

Twenty randomly selected PD outpatients, 15 men and 5 women, average age of 71.25 years, mean disease duration of 5.28 years, and Hoehn–Yahr (Hoehn and Yahr, 1967) range of 2.5–4, participated in the study. Participants were on their regular anti-Parkinsonian medication. Patients with considerable visual deficit not compensated by correction, ocular movement dysfunction and gait disturbances due to neuromuscular diseases were excluded. Disease-related disability was assessed using the Hoehn and Yahr scale.

Assessment of patient ambulation was based on walking speed (meters/second) and stride length (meters) calculated directly from measurements of time and number of steps needed to complete a fixed straight track of 10 meters.

Visual cues were generated by a belt-mounted unit, housing motion sensors and digital processing components and delivered by a micro-display, as shown in Fig. 10.1. The display, attached to the eyeglasses frame, provides the patient with a virtual tiled floor in a checkerboard arrangement, responding dynamically to the patient's own motion, much like a real floor, fixed in space. The small size of the display and the see-through feature of its enclosure ensure that the field of view is effectively unobstructed.

All tests were performed at the Carmel Medical Center, Haifa, Israel, at about the same time in the afternoon. Examination comprised five stages, each consisting of the patient walking a straight track of 10 meters: baseline, online display off, online display on, residual effects and re-evaluation. At each stage, the length of time to complete the 10-meter track and the number of steps were recorded four times and averaged. At the start of each stage, the patient was instructed to "walk normally", so as to reduce the motivational factor.

Stage 1: This stage involves walking without the device for measurements of the baseline performance.

Stage 2: This stage involves walking with the device placed on the patient, but with the display turned off, in order to measure the preconditioning effect.

Stage 3: This stage involves walking with the display turned on in order to produce and measure the online effect of the virtual tiled floor. Patients were instructed to "walk on the virtual tiles", using their imagination in trying to "reach for the next tile" with each of their feet. No further restrictions, regarding stepping on tiles of a specific color or not stepping on tile edges, were imposed or suggested.

Stage 4: This stage involves walking without the device after a 15-minute break in order to measure the short-term residual effect of the visual feedback cues.

Stage 5: This stage involves a re-evaluation of the baseline performance without the device one week after the first examination in order to measure the long-term residual effect of the visual feedback cues.

The clinical test results are given in Table 10.2.

With the device mounted, but the display turned off, the patients showed 1.75% improvement in walking speed and 2.61% improvement in stride length, on average, with respect to the baseline performance. These changes represent a negligible "placebo" effect. Turning the display on, 65% of the patients improved either their walking speed or their stride length or both by over 10%, and 55% of the patients showed over 20% improvement. Following the 15-minute break, 85% of the patients showed over 10% improvement, and 65% showed over 20% improvement in either walking speed or stride length or both without the device. The differences between residual and online performance can be attributed to burdening and learning effects associated with using the device for the first time and are expected to diminish with further training. One week after participating in the first test (where the patient used the device for approximately 20 minutes), 11 of the patients returned for another "baseline" test. Of these patients, 64% showed over 10% improvement and 36% showed over 20% improvement in their baseline performance. Two of the patients reported that, a week later, they still walked on the virtual tiles in their minds.

In this study, we found that the online visual feedback-induced improvement in the walking abilities of patients of on their regular medication schedule (56% of the patients showing over 20% improvement in either walking speed or stride length) is similar to that found in our previous study (Baram *et al.*, 2002) on patients taken off their

Table 10.2. Test results for patients with PD.

Patient				Baseline		Display off		Display on		15-minute residual	
Patient	Age/Sex	DD (years)	HY	Walking speed (m/s)	Stride length (m)	Walking speed (m/s)	Stride length (m)	Walking speed (m/s)	Stride length (m)	Walking speed (m/s)	Stride length (m)
1	80 M	10	3.5	0.782	0.500	0.752	0.488	0.790	0.677	0.867	0.533
2	75 F	3	2.5	0.669	0.408	0.628	0.400	0.627	0.400	0.747	0.435
3	77 M	3	3	0.848	0.476	0.816	0.465	0.954	0.541	1.059	0.588
4	61 M	1	3	1.066	0.526	1.071	0.526	1.310	0.625	1.147	0.571
5	79 F	19	3.5	0.680	0.408	0.686	0.400	0.618	0.444	0.735	0.426
6	83 M	5	2.5	0.675	0.426	0.746	0.444	0.606	0.454	0.815	0.454
7	67 M	10	3	1.229	0.645	1.368	0.694	1.426	0.769	1.530	0.741
8	60 M	1	3	1.100	0.606	0.980	0.555	1.373	0.769	1.238	0.667
9	53 F	3	3	1.074	0.625	1.161	0.645	1.100	0.588	1.314	0.645
10	62 M	5	3	1.291	0.645	1.316	0.667	1.356	0.667	1.517	0.741
11	68 M	9	3	0.980	0.513	1.021	0.526	1.358	0.714	1.285	0.588
12	80 M	4	2.5	0.829	0.500	0.933	0.526	1.151	0.588	1.033	0.571
13	78 F	5	2.5	0.602	0.417	0.694	0.435	0.752	0.454	0.849	0.500
14	77 M	5	2.5	0.818	0.513	0.946	0.556	1.063	0.645	1.104	0.588
15	73 M	1	3	0.539	0.308	0.591	0.339	0.808	0.476	0.873	0.500
16	70 M	3.5	3.5	0.370	0.200	0.232	0.150	0.380	0.208	0.375	0.227
17	61 F	7	2.5	0.422	0.323	0.460	0.333	0.439	0.333	0.440	0.345
18	74 M	5	3	0.154	0.376	0.133	0.427	0.294	0.343	0.282	0.374
19	72 M	5	3.5	0.027	0.028	0.022	0.026	0.020	0.025	0.033	0.030
20	75 M	1	4	0.389	0.263	0.486	0.357	0.525	0.408	0.479	0.357

Note: DD: disease duration, HY – Hoehn–Yahr staging. "Display off" columns present walking speed and stride length results for patients wearing the device with the display turned off. "Display on" columns show online performance results with the display turned on. "15-minute residual" columns present the performance results after a 15-minute break when patients walked freely with the device taken off.

medication for 12 hours (57% of the patients improving by over 20%). Furthermore, the present study shows that, following VR training, PD patients experience a sustained improvement in their walking abilities. This finding is in line with a recent case study, showing a sustained improvement in a single subject following one-month training with visual cues placed on the ground.

The fact that PD patients on a regular medication schedule, trained with a visual feedback VR device for less than 20 minutes, have a persistent significant improvement in both stride length and walking speed with only a slight decline is particularly encouraging. Some patients described it as a sensation of walking on the tiles in their mind. This interesting "learning process" can be supported by animal model studies, suggesting that the basal ganglia provide an internal non-specific cue to trigger movements and imply that PD involves a deficiency in this cueing. However, PD patients typically rely upon external visual cues. It is possible that such external cues perpetuate imprinting of internal cues, forming a "cue memory" that helps maintain certain motor skills.

Since PD is a chronic progressive disease, we suggest that the visual feedback VR technique should be examined as a long-term procedure in addition to physical therapy to maintain gait and postural performances in PD patients.

10.5 Discussion

Visual cues, long known to affect gait patterns, provide a channel for non-pharmacological therapeutic intervention in PD. In the present study, we have demonstrated that it is possible to use virtual reality cues, superimposed on the real world, to help PD patients control their gait. In particular, we have shown that the best effect can be achieved using a closed-loop display that responds to the patient's own motion. Specifically, our device displays a virtual tiled floor, which, as a real floor, appears to be fixed in space, moving with respect to the observer only when he or she does. In contrast, we have found that an open-loop, non-adaptive cue display causes adverse effects, such as dizziness, nausea, or even freezing. Predictive sensory cues, such as serial transverse lines, may trigger step transitions during walking. Furthermore, transverse lines may help a PD patient scale stride length, which is abnormally short otherwise, possibly due to underestimated muscle activity required to produce sufficient movement. We suggest that our closed-loop virtual display system produces the same effects as a real tiled floor. We further suggest that the failure of the open-loop virtual display system to produce satisfactory gait control in most patients is caused by the perpetual motion of the image, which, being unaffected by the patient, is eventually neglected. If, on the other hand, the patient attempts to keep attention on the perpetual floor motion, the relevant optokinetic system gets continuously stimulated, causing dizziness. In addition, the perpetual floor motion in the open-loop system does not forgive the patient for deviating from the set speed, resulting in a sense of confusion and being 'out of sync'. In contrast, the patient using the closed-loop system has, by body motion, full control of the relative motion of the virtual floor, which, by biofeedback, helps regulate gait. A previous study (Hanakawa *et al.*, 1999) explored the functional neuroanatomy that underlines visually guided gait in PD patients. They found that the right lateral pre-motor cortex, which is mainly regulated by cerebellar inputs, was activated to a greater extent in PD patients than in age-matched healthy individuals by visual transverse lines. On the other hand, the healthy individuals activated mainly the SMA, which was underactivated in PD patients. It appears, then, that visually enhanced gait employs different brain pathways in PD patients compared to healthy individuals, bypassing the impaired SMA function. The proposed augmented reality apparatus may serve as a mechanism for activating this alternative pathway. Our study reveals that the gait parameters, which are most sensitive to anti-Parkinson medication and are improved by pallidotomy (brain surgery), namely, walking speed and stride length, can also be manipulated, to a similar extent and without some adverse effects, by a closed-loop display of virtual visual cues. The full extent of this approach within a comprehensive diagnosis and care program for movement disorder patients is proposed for further investigation. We note, in closing, that beyond the seminal study (Baram *et al.*, 2002), the studies reported above point out both short-term residual signatures following virtual visual patient training, which have also surfaced in our other studies (see subsequent chapters), and EEG recordings indicating cortical activity matching gait entrainment.

Chapter 11

Auditory Feedback for Gait Improvement in Parkinson's Disease

11.1 Introduction

The effects of sensory cues on gait in Parkinson's Disease (PD) patients have been noted as early as half a century ago (Martin, 1967) and studied extensively in the following years (e.g., Dunne *et al.*, 1987; Bagley *et al.*, 1991; Morris *et al.*, 1994; Azulay *et al.*, 1999). However, the unique closed-loop visual feedback aspect of motion over the earth-stationary markings provided in that early study has been overlooked and largely misunderstood in early technological attempts to generate cuing by externally enforcing constant visual motion in an open-loop fashion (e.g., Prothero, 1993). As analytically demonstrated (Baram, 1999) and clinically tested in comparison to open-loop visual cuing (Baram *et al.*, 2002), the fuller extent of closed-loop visual feedback, imitating environmental effects, requires more advanced technological means, involving motion sensors and real-time computation. Open-loop, metronome-like, rhythmic auditory cues, externally enforcing a gait pattern in a feedforward fashion, have been reported to have certain positive effects in PD (e.g., Kritikos *et al.*, 1995; Thaut *et al.*, 1996; Howe *et al.*, 2003; Nieuwboer *et al.*, 2007; Delval *et al.*, 2014). While both visual (Baram and Miller, 2006) and auditory (Baram and Miller, 2007) closed-loop feedback effects were applied in studies of gait improvement in multiple sclerosis (MS) patients, and a more recent study (Espay *et al.*, 2010) has examined the combined effects of closed-loop visual and auditory feedback on gait in PD, the isolated effects of closed-loop auditory feedback in PD do not appear to have been previously reported. It is the purpose of the present study to close that gap and examine the effects of closed-loop auditory feedback on PD gait in comparison to the effects of open-loop auditory cuing (Baram *et al.*, 2016).

11.2 Methods

11.2.1 *Subjects*

Sixteen randomly selected patients with PD, average age of 69.94 ± 7.85 years, average disease duration of 6.12 ± 4.69 years, and average Hoehn–Yahr staging (Hoehn and Yahr, 1967) of 2.53 ± 0.56 participated. The study, performed at the Rambam and the Carmel medical centers in Haifa, Israel, was approved by the respective institutional Helsinki committees, and informed consent was obtained from each of the participants once the nature of the procedures had been fully explained. Participants were on their regular anti-Parkinsonian medication. Patients with considerable hearing deficits not compensated by correction and gait disturbances due to neuromuscular diseases were excluded.

11.2.2 *Assessment of performance*

The assessment of patient performance was based on walking speed (meters/second) and stride length (meters) calculated directly from measurements of time and number of steps needed to complete a fixed straight track of 10 meters.

11.2.3 *Closed-loop auditory feedback apparatus*

A clicking sound is generated by a belt-mounted cellphone-size unit, housing motion sensors and digital processing components in response to every step taken by the patient and delivered to the patient by earphones. This allows the patient to maintain a steady, balanced gait, controlling the auditory cue by changing the gait pattern and correcting any deviation from the intended auditory cue by correcting the gait pattern. It stands in contrast to metronome-like devices, which externally enforce a rhythmic cue without providing any feedback information on performance to the patient who, by "falling out of sync", may lose confidence and balance.

All tests were performed at about the same time in the morning. Examination comprised four stages, each consisting of the patient walking a straight track of 10 meters: baseline, display off, display on, and residual effects. At each stage, the length of time to complete the 10-meter track and the number of steps were recorded four times and averaged. At the start of each stage, the patient was instructed to attempt to produce a rhythmic auditory cue by maintaining a steady, balanced gait:

Stage 1: This stage involves walking without the device for measurements of the baseline performance.

Stage 2: This stage involves walking with the device turned off and placed on the patient in order to measure the preconditioning effects.

Stage 3: This stage involves walking with the device turned on in order to produce and measure the online effect of the device. Patients were instructed to walk so as to generate a rhythmic sound with their steps. No further restrictions were imposed or suggested.

Stage 4: This stage involves walking without the device after a 15-minute break in order to measure the short-term residual effect of the closed-loop auditory feedback cues.

11.3 Results

The clinical test results are given in Table 11.1.

We compared the performance with the device turned on with the baseline performance and the performance with the device turned off. We then compared the short-term residual performance with the baseline performance.

Putting the device on the patient with the device turned off showed an average increase of $2.02\% \pm 5.44\%$ in the walking speed and an average decrease of $-0.6\% \pm 1.99\%$ in the stride length with respect to the baseline.

Walking with the device mounted and turned on, the patients showed, on average, $12.37\% \pm 18.37\%$ improvement in walking speed and $4.30\% \pm 3.64\%$ improvement in stride length with respect to

Table 11.1. Test results for patients with PD.

Patient				Baseline		Device off		Device on		15-minute residual	
Patient	Age/Sex	DD	HY	Walking speed (m/s)	Stride length (m)	Walking speed (m/s)	Stride length (m)	Walking speed (m/s)	Stride length (m)	Walking speed (m/s)	Stride length (m)
1 AW	72 M	9	3	0.782	0.425	0.797	0.417	0.835	0.435	0.903	0.465
2 YT	78 M	15	1.5	0.999	0.588	0.997	0.571	1.191	0.714	0.859	0.513
3 SM	65 M	16	2.5	1.174	0.541	1.358	0.555	1.241	0.541	1.323	0.588
4 BD	58 M	6	2	1.055	0.645	1.013	0.645	1.811	0.667	0.855	0.476
5 GT	62 M	5	2	1.350	0.645	1.316	0.667	1.338	0.667	1.375	0.667
6 AZ	70 M	2	2.5	0.782	0.455	0.785	0.455	0.891	0.513	0.925	0.519
7 BW	63 F	4	1.5	1.071	0.588	1.131	0.588	1.186	0.625	1.200	0.625
8 HM	78 F	5	2.5	0.737	0.444	0.785	0.476	0.825	0.488	0.828	0.476
9 LO	77 M	5	2.5	1.007	0.571	1.052	0.588	1.206	0.667	1.104	0.625
10 AF	73 M	1	3	0.822	0.454	0.842	0.465	0.927	0.500	0.921	0.500
11 YK	80 M	10	3.5	0.902	0.556	0.975	0.588	0.934	0.588	0.929	0.588
12 AS	77 M	3	3	0.877	0.500	0.893	0.500	0.914	0.526	0.941	0.526
13 SB	61 M	1	2.5	1.148	0.556	1.095	0.526	1.120	0.556	1.133	0.556
14 RA	67 M	10	3	1.305	0.645	1.285	0.645	1.340	0.714	1.475	0.667
15 ZB	83 M	5	2.5	0.823	0.454	0.777	0.435	0.837	0.465	0.806	0.444
16 YK	60 M	1	3	1.125	0.625	1.168	0.667	1.342	0.714	1.270	0.690

Notes: DD: disease duration (years), HY: Hoehn–Yahr staging. "Baseline" columns present the walking speed and stride length results for patients before any use of the auditory feedback device. "Device off" columns present the walking speed and stride length results for patients wearing the device with the device turned off. "Device on" columns show the online performance results with the device turned on. "15-minute residual" columns present the performance results after a 15-minute break when patients walked freely with the device taken off. Patients 2 and 4 went into "off" in residual tests (by their own testimony).

the baseline performance. The average improvement with respect to walking with the device mounted but turned off was 10.72% ± 19.53% in walking speed and 6.77% ± 6.57% in stride length.

Noting that patients 2 and 4, by their own testimony, went into "off" during their short-term residual testing, and removing these patients' short-term residual performance results, we obtain 9.09% ± 6.34% improvement in walking speed and 6.52% ± 4.36% improvement in stride length compared to the baseline performance.

Device-on performance can be seen to improve walking speed in 87.5% of the patients and stride length in 81.25% of the patients compared to the device-off performance. Average short-term residual performance can be seen to represent improved walking speed in 75% of the patients and improved stride length in 68.75% of the patients compared to the baseline performance. Taking patients 2 and 4 (who, as noted, went into "off" during their short-term residual testing) off the count here increases the improvement results in both walking speed and stride length to 85.71%.

We note that there is a high correlation between walking speed and stride length. Indeed, employing the results presented in Table 11.1, we find that the correlation coefficient between the two measures in the baseline performance is 0.8672. The online (device-on) correlation coefficient between walking speed and stride length is 0.7714, while the residual correlation coefficient between the two measures is 0.8848. The strong positive correlation between walking speed and stride length, and vice versa, raises the question of whether there is also a strong correlation between improvement in walking speed and improvement in stride length. Employing Table 11.1 again, we find that the correlation coefficient between the online improvement in walking speed and the online improvement in stride length compared to the baseline performance is 0.1819. Similarly, the correlation coefficient between the online improvement in walking speed and the online improvement in stride length compared to the device-off performance is 0.2182. On the other hand, the correlation coefficient between the residual improvement in walking speed and the residual improvement in stride length compared to the baseline performance is 0.9141. Clearly, residual improvement shows considerably higher correlation between walking speed and stride length. The reason for the large discrepancy between the online and residual results in this respect is left for further research.

While we have reported changes in average walking speed and stride length, average changes in cadence could also be calculated by simply dividing average walking speed by average stride length. We note, however, that the interpretation of average change in cadence would be highly case-dependent. For instance, high average cadence may imply severe gait impairment in PD patients, on the one hand, or efficient locomotion in healthy individuals or slightly impaired patients, on the other. A more informative measure would be the rate of change in cadence, indicative of gait impairment severity in PD, which, however, was not measured in this preliminary study due to a lack of suitable equipment.

11.4 Discussion

Previously reported application of open-loop (metronome-like) auditory cuing has been found to increase or decrease walking speed in PD patients according to changes in the cue frequency above or below the preferred pace, respectively (Howe *et al.*, 2003). At the same time, stride length has not been found to be significantly affected by such changes or by open-loop auditory cuing altogether (Howe *et al.*, 2003; Delval *et al.*, 2014). At the preferred pace, open-loop auditory cuing has been reported to result in 4.2% average improvement in the PG score, combining walking speed and stride length, among a variety of posture and gait measures (Nieuwboer *et al.*, 2007). Our present study has found that closed-loop auditory feedback, releasing the patient from externally enforced cuing, results not only in relatively high average improvement in walking speed (12.4%) but also in non-negligible average improvement in stride length (4.3%). Notably, the participants in the present study were non-freezers, and consequently, the effects of closed-loop auditory feedback on freezing have not been addressed here.

The present study has found a pronounced short-term residual effect of closed-loop auditory feedback on both walking speed (9.1%) and stride length (6.5%). In contrast, open-loop auditory cuing has been found to have no functional carryover effects (Nieuwboer *et al.*, 2007). A study of combined visual and auditory closed-loop feedback (Espay *et al.*, 2010) has found long-term residual improvement in both walking speed and stride length. We suggest that the short-term residual effect of training with closed-loop auditory feedback, found in the present study, may present a gait improvement modality, which should be studied further, along with its long-term effects. While the discrepancy we found between the highly correlated walking-speed and stride-length improvements in residual performance and the poor correlation between the improvements in these measures in online performance may be generally explained by the discrepancy between short-term and long-term learning and memory effects, we suggest this discrepancy for more detailed future research.

Finally, we note that, while a large majority of the patients tested have shown improvement in gait parameters as a result of closed-loop auditory feedback, the results also show relatively high deviation in the level of improvement among patients, which indicates that different patients may benefit differently from such interventions. Another interpretation may call for more extensive instructional patient preparation and training, which have not been attempted in the present study to a significant degree.

Chapter 12

At-Home Audio-Visual Training by Parkinson's Patients

12.1 Introduction

Gait impairment in the form of shuffling, short steps, or freezing often occurs during the course of Parkinson's disease (PD), generating substantial disability (Giladi *et al.*, 2001). It is well recognized that auditory, visual, or tactile stimuli can improve gait in patients with PD (Lim *et al.*, 2005). As a result, devices that exploit the sensory cueing-related modification of gait in patients with PD have been proposed with limited success for ongoing use (Baram *et al.*, 2002; Ferrarini *et al.*, 2008). These devices employ open-loop strategies, i.e., they impose on the patient a sensory signal, generated by an external source, that is not affected by the patient's own motion. Examples are fixed-velocity visual cues (Prothero, 1993) or rhythmic auditory cues (Thaut *et al.*, 1996). While open-loop stimulation may offer some benefit to some patients (Baram *et al.*, 2002; Nieuwboer *et al.*, 2007), such cues were found to cause confusion and exacerbate freezing of gait in others (Baram *et al.*, 2002). Indeed, it is well known that the stabilization of open-loop dynamic systems, which are inherently unstable, can only be achieved by feedback, i.e., closed-loop, systems (Athans and Falb, 1966; Kuo, 1981). The key role played by sensory feedback in gait stabilization and control, i.e., where the feedback signal is generated by the patient's own motion, has been demonstrated by the gait improvements in patients with PD when they are provided with earth-stationary visual markings (Martin, 1967; Baram, 1999). At the same time, a clinical study addressing the role of visual feedback in maintaining postural balance in stance demonstrated that feedback increases postural stability in standing tasks, indicating a more prominent role for closed-loop (feedback) control over open-loop (feed-forward) control (Rougier, 1999).

The development of a wearable virtual reality device driven by inertial sensors that delivers earth-stationary visual feedback cues has provided an opportunity to assess a closed-loop system of sensory (visual-auditory) feedback stimuli to aid dysfunctional gait in patients with PD. We sought to examine the effect of a wearable, closed-loop, visual-auditory cueing system on Parkinsonian gait in the "off" state when the magnitude of gait impairment is expected to be largest. Preliminary clinical studies (Baram *et al.*, 2002; Baram and Miller, 2006) have demonstrated that the closed-loop feedback device significantly improves gait speed while in use and appears to induce a residual benefit for several days by dynamically delivering a tiled floor pattern adapted to the patient's own motion. It is believed that the optical flow provides a reassuring feedback effect, informing the patient of motion being performed and matching visual feedback from foot-tile matching to auditory feedback from clicking obtained for each step. These elements help produce a balanced rhythmic walk. In contrast to the open-loop strategies, this device also appears to produce and maintain a high degree of patient confidence, comfort, and safety (Baram *et al.*, 2002). To the best of our knowledge, this study (Espay *et al.*, 2010) is the first to assess the feasibility and effects of at-home training with a closed-loop sensory-feedback device in patients with PD. Finally, we note that this report includes some medical terminology, which may be skipped by the less familiar reader.

12.2 Methods

12.2.1 *Subjects*

We recruited consecutive, consenting patients with PD who had off-state gait shuffling, festination, and/or freezing and a score >1 on the Unified Parkinson Disease Rating Scale (UPDRS) III item 29 and who did not require continuous ambulatory aid from the Movement Disorders Center at the University of Cincinnati. Patients spent over 3 hours/day in the off state. We did not study patients with primary on-state freezing to maximize homogeneity of the target population. Patients had been on a stable dose of antiparkinsonian medications and required no medication adjustments during the course of the study. The following exclusion criteria were used: (1) prior functional neurosurgery; (2) musculoskeletal disorders, such as severe arthritis, knee surgery, hip surgery, or any other condition that the investigators determined would impair assessment of gait; (3) history of stroke; (4) cerebellar, vestibular, or sensory ataxia; (5) Mini-Mental State Examination score smaller than 24.

12.2.2 *Visual-auditory walker*

The device is composed of a small measurement-computation unit attached to the patient's clothing, a head-mounted microdisplay, and earphones (Fig. 12.1). The measurement-computation unit is composed of a multiaxial accelerometer, a compass, and a microcontroller. The apparatus, operating in an adaptive closed-loop mode, displays a life-size virtual checkerboard-tiled floor superimposed on the real world with specialized see-through glasses. The closed-loop (or feedback) concept (as opposed to open loop) implies that the speed of the cues is not externally set but, rather, is an outcome

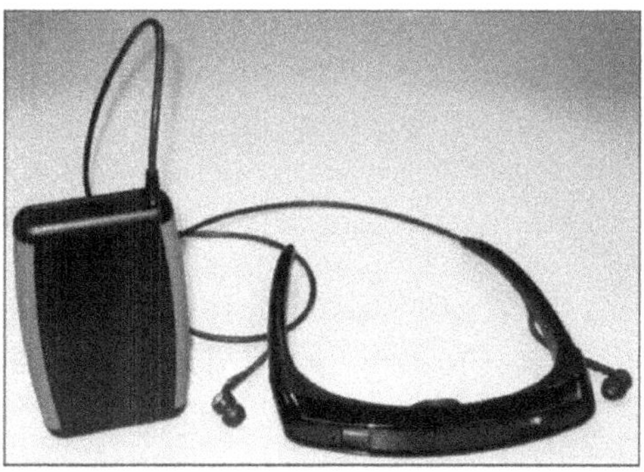

Figure 12.1. Virtual (augmented) reality device used in this study containing a built-in LCD screen, which projects floor tiles when subjects are moving, and earphones that sound step-matched cues, as determined by a connected sensor strapped to the belt.

(and indeed a duplicate) of the walking speed of the user. The visual effect is the same as that created by walking over earth-stationary cues (i.e., a real tiled floor). Similarly, the rhythm of the auditory cue is determined by the rhythm of the steps, not vice versa. The user then regulates the gait pattern to create a constant optical flow and a rhythmic auditory cue. The virtual (augmented) floor responds dynamically to the patient's own motion and "moves" toward him or her at the speed set by the user, as measured by the accelerometer. The tiled floor acts as a moving visual display whose speed is generated in a natural feedback fashion by the patient's own motion, much like earth-stationary visual cues. The grid allows patients to step on the tiles with long strides as they walk, though they do not enlarge or modify based on the previous step length. A steady gait synchronizes the patient's own steps with the virtual tiles and the auditory cues, thus "rewarding" the patient for making the effort. Subjects can view real-world obstacles without difficulty. Subjects with corrective lenses are able to use the device without difficulty. Additional auditory feedback from the patient's own steps is provided through earphones. Subjects hear the auditory cue produced by the steps through the earphones. The auditory feedback is continuous so long as patients are walking steadily, producing a rhythm they hear based on their gait pattern.

12.2.3 *Gait system*

The validated gait analysis system (GAITRite, CIR Systems, Inc.; Havertown, Pennsylvania) used in this study consists of a 4-meter electronic walkway that contains six sensor pads encapsulated in a roll-up carpet to produce an active area 24 inches (61 cm) wide and 144 inches (366 cm) long (Webster *et al.*, 2005). In this arrangement, the active area is a grid (48 sensors by 288 sensors placed on 0.5-inch [1.27-cm] centers) totaling 13,824 sensors. The walkway is portable, can be laid over any flat surface, requires minimal setup and test time, and requires no placement of any devices on the patient. The application software controls the functionality of the walkway, processes the raw data into footfall patterns, and computes temporal (timing) and spatial (distance) gait parameters, including stride length, cadence, and velocity.

12.2.4 *Study evaluations*

Subjects were asked to walk at their usual pace on a defined 10-meter path encompassing the GAITRite walkway beginning and ending at the same point under five conditions: (1) no sensory feedback (no device, baseline); (2) visual-auditory device positioned but deactivated (device off); (3) visual-auditory device positioned with visual-only feedback activated; (4) visual and auditory feedback activated; (5) again with no sensory feedback (device off, immediate residual effect). Each condition was repeated three times, and the average of these was used for analysis. Due to the nature of the intervention, the order of the assessments was not randomized, and subject blinding was unattainable (signal noise in device would have worsened gait in any control group), but a comparison of conditions 1 (no device) and 2 (device off) served to determine whether a placebo-like response could be elicited when the device was not yet providing visual-auditory feedback. Subjects were assessed at two time points: baseline (visit 1) and after 2 weeks of at-home use (visit 2). In between these in-clinic evaluations, subjects were instructed to use the visual-auditory device while walking for no less than two 30-minute periods a day, regardless of their medication state. We emphasized use during periods of expected off-related gait worsening. Subjects were not restricted in their choice of walking inside or outside and were encouraged to incorporate training with their regular activities. Since subjects were recruited on the basis of their off-related gait impairment, all study assessments took place during the "practically defined off period", i.e., in the morning at least 12 hours after the last dose of any anti-Parkinsonian medication. We administered a standard motor examination with the UPDRS III (Fahn and Elton, 1989) and asked subjects to complete the self-administered Freezing of Gait Questionnaire (FOGQ, range 0–24; lower is better) (Giladi *et al.*, 2000) to ascertain their motor and gait-specific function.

12.2.5 *Data analysis*

We used the primary outcome measures (gait velocity, stride length, and cadence) to assess the effect of the device under various conditions. To accomplish this, we evaluated the effects of the various conditions (no device, device without feedback, visual-only feedback, visual-auditory feedback, and no device for immediate residual) on each outcome variable using repeated measures analysis of variance followed by *post hoc* multiple comparisons with Bonferroni adjustment for each visit separately. The multiple comparisons were for each of the conditions versus the baseline condition (condition 1). We examined the effect of training on each outcome using paired t-tests for each condition separately. Secondary outcome measures included changes in total FOGQ score, UPDRS III score, axial UPDRS III score, and freezing frequency between baseline and follow-up. We calculated the axial UPDRS III subscore by adding motor items 18–19 and 27–30, as described in a previous study (Espay *et al.*, 2005), to determine whether, in addition to gait, other axial motor features of PD were affected by the device. We used paired t-tests to compare the secondary outcomes between the visits. As an exploratory analysis, we used multiple regression analysis to assess the effect of baseline UPDRS on the absolute change in velocity and stride length between condition 5 at baseline and follow-up, adjusting for disease duration. The significance criterion for the regression analysis was kept at 0.05. We considered p-values less than 5% significant.

12.3 Results

12.3.1 *Clinical descriptions*

Fifteen patients were enrolled, but two did not feel comfortable using the glasses (because they were "clunky" or "embarrassing to use in public") and did not train at home as instructed. The patients (six males, 73.3 ± 11.7 years old, disease duration: 12.1 ± 4.2 years, levodopa equivalents: $1{,}413.8 \pm 786.2$ mg) had a UPDRS III score (off-state during both assessments) of 28.1 ± 11.1 at the first assessment and 24.1 ± 9.1 at the second assessment (all data presented as mean $\pm$ standard deviation unless otherwise noted). Patients reported varying preferences for sensory feedback: Four subjects favored the combination of auditory and visual feedback, three subjects preferred auditory stimulation alone, three preferred visual feedback alone, and two preferred neither (data missing for one subject). The gait parameters obtained for each condition with the gait analysis system are shown in Table 12.1.

Table 12.1. Gait parameters per condition (data presented as mean ± standard deviation).

Condition	Velocity (cm/s)	Stride length (cm)	Cadence (steps/min)
No Device (baseline)			
Visit 1	64.2 ± 18.8	75.0 ± 15.4	104.4 ± 22.0
Visit 2	69.2 ± 22.8	81.1 ± 19.9	108.9 ± 23.3
%Δ	7.7 ± 14.8	8.1 ± 15.1	7.1 ± 23.0
Device Without Feedback (off)			
Visit 1	60.6 ± 18.3	73.9 ± 15.9	117.1 ± 27.3
Visit 2	71.5 ± 23.4	86.1 ± 23.7	114.8 ± 43.6
%Δ	18.7 ± 23.2	16.6 ± 19.4	-3.7 ± 15.6
Visual-Only Feedback			
Visit 1	56.4 ± 22.0	69.6 ± 20.5	105.3 ± 26.2
Visit 2	69.5 ± 26.8	80.4 ± 22.6	109.2 ± 24.5
%Δ	$26.5 \pm 29.1^{*}$	19.2 ± 28.4	5.9 ± 20.7
Visual-Auditory Feedback			
Visit 1	61.6 ± 20.1	74.3 ± 16.4	105.4 ± 32.0
Visit 2	72.6 ± 26.5	84.0 ± 18.5	110.6 ± 41.8
%Δ	$17.1 \pm 20.4^{\dagger}$	$13.6 \pm 12.7^{\dagger}$	3.3 ± 12.6
No Device (immediate residual)			
Visit 1	64.5 ± 21.4	79.0 ± 20.3	107.4 ± 42.3
Visit 2	75.4 ± 21.5	88.8 ± 17.7	114.6 ± 25.8
%Δ	$18.7 \pm 14.4^{\ddagger}$	$14.9 \pm 15.2^{\ddagger}$	14.2 ± 35.0

Notes: *Significant difference between visits $p < 0.05$; †significant difference between visits $p < 0.01$; ‡significant difference between visits $p < 0.001$; Diff = Visit 2 − Visit 1, %Δ = percent change ((Diff/Visit 2) × 100).

12.3.2 *Condition effect by visit*

At visit 1, a significant effect of the device on gait velocity was noted under various conditions ($p = 0.006$). The multiple pairwise comparisons revealed that condition 2 was not significantly different from condition 1 ($p = 0.39$). However, condition 3 was significantly different from condition 1 ($p = 0.008$). At visit 2, the effect of the device on gait velocity for various conditions was found to be statistically significant ($p = 0.03$). In *post hoc* pairwise comparisons, condition 5 was significantly different from condition 1 ($p = 0.01$), showing that a residual effect of the device on gait velocity existed. No significant changes in mean stride length were noted at different conditions at visit 1 ($p = 0.18$). However, a significant improvement in stride length was found at visit 2. Only condition 5 was found to be significant compared to condition 1 ($p = 0.008$). Cadence was not significantly different under any condition for any visit. In summary, a condition effect on gait velocity as well as stride was found. The positive effects of the device on gait velocity and stride were mainly observed at the second visit between condition 5 and condition 1.

12.3.3 *Training effect for condition* 1

A trend for improvement in velocity (64.2 ± 18.8 cm/s to 69.2 ± 22.8 cm/s, $p = 0.05$) and stride length (75.0 ± 15.4 cm to 81.1 ± 19.9 cm, $p = 0.08$) was noted. Changes in cadence (104.4 ± 22.0 steps/minute to 108.9 ± 23.3 steps/minute) were not significant. Gait velocity improved by at least 10% in six subjects and by more than 20% in two subjects. Stride length showed similarly graded responses (Fig. 12.2).

12.3.4 *Training effect for condition* 2

A significant effect of training was noted on gait velocity ($p = 0.02$) and stride ($p = 0.02$) but not on cadence ($p = 0.70$) for condition 2. This result indicates that improved training effects were obtained with the use of the device.

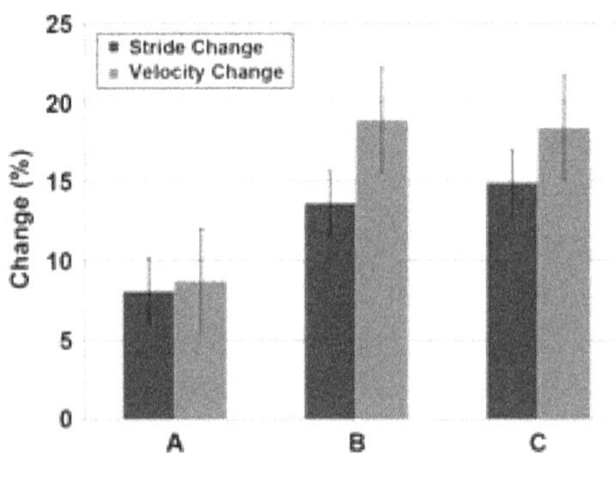

Figure 12.2. Percent change of velocity and stride length due to training effect (A = condition 1, day 1 vs. day 14, no sensory feedback), combined device and training effects (B = condition 4, day 1 vs. day 14, sensory feedback), and combined training and immediate residual effects (C = condition 5, day 1 vs. day 14, no sensory feedback). Vertical bars denote standard error of the mean.

12.3.5 *Training effect for condition 3*

Under condition 3, a significant improvement in gait velocity was noted ($p = 0.03$). Also, a trend toward improvement in stride length was found ($p = 0.06$). However, again, no change was noted in cadence for condition 3 ($p = 0.51$).

12.3.6 *Training effect for condition 4*

An improvement was found in gait velocity (61.6 ± 20.1 cm/s to 72.6 ± 26.5 cm/s, $p = 0.006$) and stride length (74.3 ± 16.4 cm to 84.0 ± 18.5 cm, $p = 0.004$). The improvement in cadence was not significant. Gait velocity improved by at least 20% in eight subjects and by more than 30% in five subjects. Stride length improved by at least 10% in nine subjects and by more than 20% in five subjects (Fig. 12.2).

12.3.7 *Training effect for condition 5*

Marked improvement was found in gait velocity (64.5 ± 21.4 cm/s to 75.4 ± 21.5 cm/s, $p < 0.001$) and stride length (79.0 ± 20.3 cm to 88.8 ± 17.7 cm, $p = 0.003$), but not cadence. Gait velocity improved by at least 10% in nine subjects and by more than 30% in three subjects. Stride length improved by more than 10% in nine subjects and by more than 20% in five subjects (Fig. 12.2).

12.3.8 *Training effect summary*

The results indicate that the training effect became more magnified under condition 4 and condition 5 for gait velocity as well as stride compared to condition 1. Also, a significant improvement in velocity and stride was found for condition 5 compared to condition 1, especially at visit 2. Thus, an effect of training and condition on velocity and stride exists.

12.3.9 *Unified Parkinson disease rating scale and freezing of gait questionnaire*

Higher baseline UPDRS scores predicted residual velocity and stride length improvement ($p = 0.002$) adjusted for disease duration, indicating that patients with more advanced disease showed greater benefit. The FOGQ improved from 14.2 ± 1.9 to 12.4 ± 2.5 ($p = 0.02$). A trend toward decreasing frequency of freezing was noted from visit 1 (2.92 episodes/day) to visit 2 (2.54 episodes/day, $p = 0.09$). The UPDRS III at visit 2 was 24.1 ± 9.1, lower than at visit 1 (28.1 ± 11.1) despite no changes in medications between the two off-state assessments ($p = 0.009$). A trend toward a reduction in the axial UPDRS scores was found at visit 2 ($p = 0.07$).

12.4 Discussion

Training with augmented reality visual-auditory feedback cues improved gait velocity by 18–20% in the immediate carryover state in a cohort of patients with PD who had substantial off-state gait impairment following a 2-week at-home period of self-training. Contrary to our expectations, the benefits were not observed within each visit, immediately upon device activation or at the initial post-training evaluation before device activation, but as a robust immediate residual or carryover effect at the end of the last assessment. In fact, nearly 70% of the subjects showed residual improvement of at least 20% in velocity, stride length, or both (with almost 40% of subjects improving by more than 30% in those measurements), changes that were significant despite high gait variability in response (as shown by large standard deviations). We speculate that the hardware required for the delivery of augmented sensory cues hindered the within-visit device-on response (by partially obstructing the visual field, adding discomfort with the belt sensor, etc.), which explains the greater benefits seen immediately after device removal compared to when the device was in use. This effect is expected to diminish with regular training and use of the device. Furthermore, improvement in the FOGQ, an instrument highly correlated with other measures of quality of life in PD (Moore *et al.*, 2007), suggests that the changes in gait were clinically significant. Patients with more advanced disease showed greater benefit after using the device, as greater baseline disability would provide the opportunity for a proportionally larger magnitude of benefit, whereas small baseline disability makes it harder for the intervention to show any beneficial effects. The reduction in at-home freezing severity noted in the FOGQ, although small in magnitude, is particularly relevant because of the pharmacoresistance of freezing and its association with an increased rate of falls (Gray and Hildebrand, 2000). A possible mechanism for the reduction in freezing, we speculate, is the generation of cortical reorganization or neuroplasticity, which may alter or circumvent the neuronal pathways responsible for this phenomenon. Finally, the improvement in velocity documented here was dependent on increases in stride length at the relative expense of cadence. The lack of change in cadence is a reflection of the decrease in step frequency seen when slow to medium walking velocities are replaced with higher walking velocities resulting from increased stride length (Morris *et al.*, 1994).

This is the first study examining the effects of a virtual (or augmented) reality device for at-home use aimed at improving gait in patients with PD. A prior randomized crossover trial of an external cue training program provided at home by physiotherapists demonstrated small but significant improvements in the Posture and Gait Scale score, a composite of related UPDRS items (13–15 and 29–30) (Nieuwboer *et al.*, 2007). While that study was the first controlled trial that used a cueing device at home, it relied on an open-loop system with cueing parameters that needed separate adjustments to increase stride length and avoid festination and a training program led by a rehabilitation specialist. These features hinder its feasibility for wider application. Further, 67% of the study subjects preferred auditory cueing over visual cueing (light flashes delivered through a light-emitting diode attached to a pair of glasses). Auditory cues may be less effective than visual cues (Cubo *et al.*, 2004).

The rehabilitation value of visual cueing, among other sensory feedback, was entered into the literature by an innovative neurologist (Martin, 1967). He suggested that the placement of visual cues perpendicular to the direction of gait spaced one step length apart was most effective in improving gait in patients with PD. Many single-session studies have confirmed the benefit of visual cues (high-contrast transverse floor lines) (Morris *et al.*, 1996; Lewis *et al.*, 2000; Jiang and Norman, 2006; Van Wegen *et al.*, 2006). However, these open-loop feedback systems may not have long-term effects (Rochester *et al.*, 2007) unless a dedicated physical training program is established, as demonstrated in a patient trained to walk on floor cues reaching 120% of the uncued stride length over a 1-month period (Sidaway *et al.*, 2006). Closed-loop feedback systems do not require training program monitoring and may lead to long-term motor skills learning and enhancement of adaptive cerebral plasticity, particularly with the use of visual information on which the generation of motor plans is dependent in patients with PD (Helmich *et al.*, 2007). While an earlier study found closed-loop systems to be safer for patients (Baram *et al.*, 2002), selected patients may benefit from open-loop systems as well (Prothero, 1993). It is important to emphasize that adherence to a daily training program (a minimum of 30 minutes of use twice per day) was critical to realizing the value of the closed-loop feedback system evaluated in this study because immediate effects were not significant. The training effort needed at the initial visit was minimal, and patients encountered little difficulty in using this device at home without further guidance.

12.4.1 *Limitations of this study*

We must point out several limitations in our study. First, the absence of a control group was unavoidable given the device's lack of a "neutral mode" (e.g., white noise and nontiled visual feedback), which could have served as an "active" placebo. Also, due to safety reasons, we could not ethically justify a placebo arm in this at-home, unsupervised gait rehabilitation type of intervention. As such, patients were unblinded to the intervention, and a potential order effect from the study conditions could have been present. The main outcome measures relied on electronic walkway rather than examiner-generated data to compensate for these methodological limitations. The lack of within-visit improvement in our patients, possibly related to subjects' cautious approach to the study hardware, argues against a placebo response, although it does not rule out physiological test-retest variability in the gait of patients with PD. Second, although we requested and verbally confirmed compliance with the requested two 30-minute training sessions per day, we did not individually monitor device use at home. This precluded assessment of a dose-response effect and subjects' preference for isolated visual or auditory feedback. The device was not uniformly accepted, and in two of the subjects, gait actually worsened while they were using it. In a similar vein, our small study did not have enough power to identify clinical variables associated with improvement versus deterioration. Similarly, the study design precluded our ability to ascertain the effect of single versus dual sensory cueing on gait, since subjects were instructed to use both visual and auditory cueing (though personal preference forced some to rely on only one or none). These issues, along with the decrease in gait performance with device use at baseline that was contrary to what was found in other device-based interventions, may cause concern regarding the safe use of the device in a therapist-free intervention, compromised compliance, and increased variability, rendering the interpretation of our results preliminary. The high variability of effect, as well as its delayed occurrence, can very likely be attributed to the learning process associated with training with and using the device, which is highly variable and depends on the patient's age and cognitive and mental capabilities. Finally, the off-state UPDRS scores were not matched at the two off-state assessments. The motor improvement at visit 2 was partly driven by a reduction in the axial scores, with which gait velocity has been found to correlate (Vieregge *et al.*, 1997) and was not explained by changes in medications or differences at the time of testing. This overall motor improvement may have acted as a confounder in the relationship between sensory cues and gait improvement. Prior studies have also shown non-pharmacological improvement in overall motor performance, such as home-based rehabilitation programs (Caglar *et al.*, 2005), and with other sensory stimulation, such as music (Pacchetti *et al.*, 1998). This is also highlighted by the discrepancy often found during routine clinic assessments, whereby patients perform better (e.g., less freezing) than what they report to experience in their homes, possibly because of the heightened attention and expectation during the increased scrutiny of clinic visits. Indeed, the testing sessions at the office likely inflated our subjects' "true" off-state motor function and raised the functional baseline against which post-intervention gait parameters were compared.

12.4.2 *Future research*

Gait impairment is a major source of disability in patients with PD for which medical and surgical strategies offer relatively little. The noninvasive and effective sensory feedback device used here may prove to be a reasonable therapeutic option for some of these patients. Further studies should ascertain the features associated with responsiveness and lack thereof, the additive effect of combined sensory cueing versus visual- and auditory-only cueing, the length of the residual benefit after training (to determine the appropriate interval that may suffice to sustain a favorable gait performance in between uses of device), the effect of this intervention on the less common but more intractable "on"-state freezing (Kompoliti *et al.*, 2000) and the benefit, if any, among post-subthalamic deep brain stimulation patients with treatment- and stimulation-refractory freezing of gait.

12.5 Conclusions

Devices using closed-loop sensory feedback, through an at-home training program, seem effective and desirable for nonpharmacologic interventions to improve walking in patients with PD. The device helped people with PD by improving gait and decreasing freezing, which are expected to have an impact on functional mobility and quality of life. Further research is necessary to quantify the duration of the residual benefit and its applicability to other clinical situations in patients with PD.

Chapter 13

Gait Improvement in Patients with Cerebral Palsy by Visual and Auditory Feedback

13.1 Introduction

Cerebral palsy (CP) is a group of perinatal disorders that present cognitive, motor, and sensory disabilities (Woollacott and Burtner, 1996). Classification is difficult as each of these disorders and disabilities appears within a wide range of independent severity. The most prominent feature is motor function disorder, associated with a general reduction in electromyogram (EMG) activity, selected muscle weakness, and afferent inhibition (Berger *et al.*, 1982; Berger, 1998). The motor impairment syndromes seen in CP are secondary to lesions or anomalies in brain development, inhibiting normal automatic and voluntary responses and allowing for the development of dysfunctional ones (Woollacott and Burtner, 1996; Berger *et al.*, 1982). Bilateral and hemiplegic spasticity give rise to difficulties in posture, balance, and gait control (Berger *et al.*, 1982; Berger, 1998; van der Heide and Hadders-Algra, 2005). Mobility-related handicaps can considerably limit daily activity, affecting individual well-being and social interaction, and therefore have a major impact on overall patient development.

The treatment of CP involves physical therapy, occupational therapy, and sports, targeted at specific conditions. Upper extremity of arm-reaching tasks was found to improve in CP patient training in a computer-generated environment (Chen *et al.*, 2007). EMG feedback was found to be potentially beneficial for gait improvement in patients with CP-related impairment, specifically dynamic equinus deformity (Dursun *et al.*, 2004) and spastic hemiplegia (Colborne *et al.*, 1994). Training with sensory feedback has been found to improve motor function. Auditory feedback, provided by a load-sensitive insole inserted in the hemiparetic side shoe, was found to have a positive effect on gait symmetry in CP (Seeger *et al.*, 1981), but maintenance strategies for this effect are yet to be determined (Seeger and Caudrey, 1983). Balance training with visual feedback was found to improve stance in hemiplegic CP with a secondary positive effect on gait (Ledebt *et al.*, 2005).

Medical testing of CP patients was facilitated by Dr. Ruben Lenger and conducted by Prof. Yoram Baram at the Sports Center for Handicapped Children in Kiryat Haim, Israel.

13.2 Materials and Methods

13.2.1 *Study population*

Ten patients complaining about gait disturbances due to CP trained with visual feedback, while 10 such patients trained with auditory feedback. While six of the patients participated in both groups, the auditory feedback training and tests were performed one week after the visual feedback training and tests so as to reduce cross-effects of the brief 20-minute training periods. Seven age-matched healthy controls trained with visual feedback, while eight such controls trained with auditory feedback. Seven of the controls participated in both groups, but the visual and the auditory training and tests were separated by 20 days. The study was approved by the Helsinki Committee of the Ministry of Health, Israel, and informed consent was obtained from a parent of each of the participants once the nature of the procedures had been fully explained.

13.2.2 *Materials*

Visual cues were generated by a VR apparatus shown on the left-hand side of Fig. 13.1. The display, attached to the eyeglasses frame,

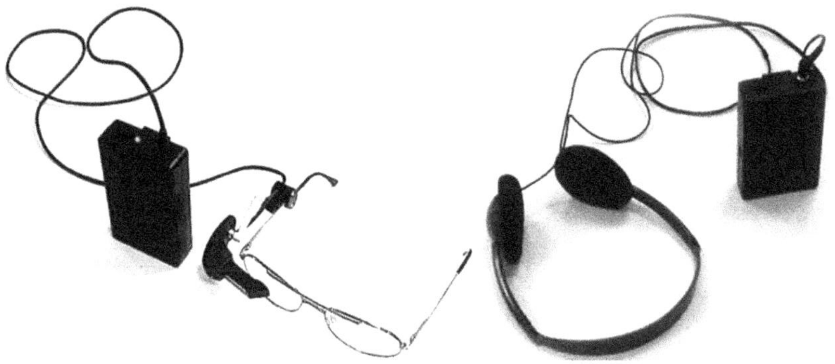

Figure 13.1. (Left) Visual feedback device used in tests; (right) auditory feedback device used in tests.

provided the patients with an image of transverse lines, responding dynamically to the patient's own motion, much like a real floor fixed in space. Auditory feedback cues were generated by a device shown on the right-hand side of Fig. 13.1, which delivers through earphones a clicking sound in response to each step. Both devices employ an inertial motion sensor and a computing unit enclosed in a belt-mounted box. Assessment of patient ambulation was based on walking speed and stride length, which are calculated directly from measurements of time and the number of steps needed to complete a fixed track.

13.2.3 *Procedure*

Examination of each patient comprised three stages:

Stage 1: Baseline walking speed and stride length without the device were measured. The patient was verbally instructed to walk along a straight track of 10 meters. The time to complete the track and the number of steps were recorded four times and averaged.

Stage 2: The device was placed on the patient and turned on. The patient was instructed to walk four times along the 10-meter track for the purpose of training. Patients training with visual feedback were asked to imagine stepping between the virtual transverse lines while walking. Patients training with auditory feedback were asked to maintain, by controlling their gait pattern, a rhythmic auditory cue.

Stage 3: The device was taken off the patient, who was given a 20-minute break. After the break, the patient was instructed to walk the 10-meter track without the device. Walking speed and stride length were recorded four times and averaged. The purpose of this stage was to measure the residual short-term therapeutic effect of training with the sensory feedback cue.

13.3 Results

13.3.1 *Visual feedback*

The test results for patient training with visual feedback are summarized in Table 13.1. The standard deviation for the four trials averaged was smaller than 10% of the mean value reported for each of the patients. The average improvement in walking speed was 21.70% ± 36.06% (the second number representing standard deviation). This result was found to be statistically significant in the sense that the hypothesis that the average improvement found corresponds to the given sample was accepted with a confidence

Table 13.1. Test results for patient training with visual feedback.

Patient			Baseline		After training		Percentage change	
Name	Sex	Age (years)	Walking speed (m/s)	Stride length (m)	Walking speed (m/s)	Stride length (m)	Walking speed (%)	Stride length (%)
1	F	10	1.285	0.555	1.265	0.571	−1.56	2.88
2	M	18	1.419	0.741	1.508	0.769	6.27	3.78
3	F	11	1.439	0.606	1.357	0.625	−5.70	3.14
4	F	7	0.44	0.328	0.648	0.385	47.27	17.38
5	F	26	0.945	0.625	1.052	0.667	10.15	6.72
6	F	9	0.401	0.27	0.369	0.263	−7.98	−2.59
7	M	6	0.81	0.377	0.939	0.476	15.93	26.26
8	F	12	0.687	0.476	1.466	0.571	113.40	19.96
9	M	15	1.224	0.588	1.513	0.645	23.61	9.69
10	F	19	1.362	0.69	1.575	0.69	15.64	0.00

level of 0.95. For participants with the baseline walking speed below the median (1.08 m/s), improvement was considerably higher (35.75% ± 47.76%) than for participants with the baseline walking speed above the median (7.65% ± 12.85%). The hypothesis was further affirmed by the fact that, while the hypotheses that the average improvements of both groups correspond to the respective samples were accepted at the 0.95 confidence level, the hypothesis that the average improvement of patients with below-median baseline performance corresponds to the above-median sample was rejected at the 0.95 confidence level. Average improvement in stride length was 8.72% ± 9.47%, with a 0.95 confidence level. For participants with baseline stride length below the median (0.57 meter), improvement was 12.78%±12.13%, and for participants with baseline stride length above the median, improvement was 4.67% ± 3.69%. While both results were confirmed at the 0.95 confidence level, the hypothesis that the average improvement of patients with below-median baseline performance corresponds to the above-median sample was rejected at the 0.95 confidence level. For participants of age below the median (11.5 years), improvement in walking speed was 9.59% ± 23.06%, while for participants of age above the median, improvement was 33.81% ± 44.96%. For participants of age below the median, improvement in stride length was 9.41% ± 11.97%, while for participants of age above the median, improvement was 8.03% ± 7.57%. However, in contrast to the dependence of the test significance on the baseline performance, no such dependence was found with respect to the education level in either the walking speed or the stride length results.

13.3.2 *Auditory feedback*

The test results for patient training with auditory feedback are summarized in Tables 13.2 and 13.4. As in the case of visual feedback, the standard deviation for the four trials averaged was smaller than 10% of the mean value reported for each of the patients. The average improvement in walking speed was 25.43% ± 28.65% with a confidence level of 0.95. For participants with the baseline walking speed below the median (0.87 m/s), improvement was considerably higher (42.79% ± 32.23%) than for participants with the baseline walking speed above the median (8.08% ± 7.39%). While the hypotheses that the average improvements of both groups correspond to the respective samples were accepted at the 0.95 confidence level, the hypothesis that the average improvement of patients with below-median baseline performance corresponds to the above-median sample was rejected at the 0.95 confidence level. Average improvement in stride length was 13.58% ± 13.10% with a confidence level of 0.95. For participants with baseline stride length below the median (0.45 meter), improvement was 13.66% ± 6.63%, and for participants with baseline stride length above the median, improvement was 14.50% ± 8.44%. While the hypotheses that the average improvements of both groups correspond to the respective samples were accepted at the 0.95 confidence level, the hypothesis that the average improvement of patients with below-median baseline performance corresponds to the above-median sample was rejected at the 0.95 confidence level. For participants of age below the median (10.5 years), improvement in walking speed was 22.83% ± 18.86%, while for participants of age above the median, improvement was 28.04% ± 38.39%. For participants of age below the median, improvement in stride length was 11.95% ± 6.94%, while for participants of age above the median, improvement was 15.21% ± 18.20%. As in the case of the visual feedback, in contrast to the dependence of the test significance on the baseline performance, no such dependence was found with respect to the education level in either the walking speed or the stride length results.

13.3.3 *Control results*

The results for the age-matched control groups training with visual and auditory feedback are given in Tables 13.2 and 13.4, respectively. For the controls who trained with visual feedback, the average improvement was −2.41% ± 9.54% in the walking speed and −2.84% ± 10.11% in the stride length. For the controls who trained with auditory feedback, the average improvement was 0.01% ± 7.73% in the walking speed and −2.03% ± 6.15% in the stride length. In addition to the small average changes in performance and the relatively large variances, it can be seen that, in both the visual and

Table 13.2. Test results for patient training with auditory feedback.

Patient			Baseline		After training		Percentage change	
Name	Sex	Age (years)	Walking speed (m/s)	Stride length (m)	Walking speed (m/s)	Stride length (m)	Walking speed (%)	Stride length (%)
1	F	10	1.224	0.513	1.342	0.555	9.64	8.19
2	M	18	1.262	0.645	1.259	0.645	6.27	0.00
3	M	6	0.851	0.417	1.266	0.5	48.77	19.90
4	F	7	0.702	0.333	0.942	0.385	34.19	15.62
5	F	26	0.892	0.555	0.914	0.588	2.47	5.95
6	F	11	1.171	0.526	1.406	0.588	20.07	11.79
7	M	5	0.841	0.351	1.006	0.4	19.62	13.96
8	M	5	1.183	0.435	1.206	0.444	1.94	2.07
9	F	12	0.543	0.426	0.629	0.476	15.84	11.74
10	F	11	0.739	0.455	1.445	0.667	95.53	46.59

Table 13.3. Test results for controls training with visual feedback.

Patient			Baseline		After training		Percentage change	
Name	Sex	Age (years)	Walking speed (m/s)	Stride length (m)	Walking speed (m/s)	Stride length (m)	Walking speed (%)	Stride length (%)
1	F	13	1.470	0.714	1.533	0.714	4.29	0
2	M	13	1.493	0.769	1.592	0.833	6.63	8.32
3	M	12	1.482	0.714	1.477	0.714	−0.34	0
4	F	12	1.460	0.728	1.487	0.769	1.85	5.63
5	F	12	1.304	0.625	0.987	0.468	−24.31	−25.12
6	M	13	1.358	0.714	1.295	0.690	−4.64	−3.36
7	M	12	1.557	0.769	1.551	0.728	−0.39	−5.33

auditory channels, about half of the controls showed a small improvement, while the others showed a small deterioration in performance. It might be concluded that, in general, neither the visual nor the auditory feedback had any measurable effect on the walking quality of the controls.

Table 13.4. Test results for control training with auditory feedback.

Patient			Baseline		After training		Percentage change	
Name	Sex	Age (years)	Walking speed (m/s)	Stride length (m)	Walking speed (m/s)	Stride length (m)	Walking speed (%)	Stride length (%)
1	F	13	1.607	0.773	1.667	0.702	3.73	−9.18
2	M	13	1.667	0.785	1.607	0.801	−3.60	2.04
3	M	12	1.75	0.817	1.690	0.716	−3.43	−12.36
4	F	12	1.833	0.728	1.667	0.714	−9.06	−1.92
5	F	12	1.547	0.678	1.778	0.714	14.93	5.31
6	M	13	1.428	0.690	1.547	0.728	8.33	5.51
7	M	12	1.667	0.741	1.547	0.690	−7.19	−5.63
8	F	14	1.667	0.769	1.607	0.769	−3.60	0

13.4 Discussion

The results of the present study support the potential role of sensory feedback-based strategies as rehabilitation modalities and substantiate their specific implementation in efforts to alleviate, improve, and restore mobility in patients with gait disturbances due to CP. For patients using visual feedback, baseline walking speed and stride length were found to be good predictors of improvement, with higher improvement associated with lower baseline performance. For patients using such cues, age was found to be a good predictor of improvement in walking speed, but not in stride length, with significantly higher improvement in walking speed found in patients of older age. For patients using auditory feedback, baseline performance was found to be a good predictor of improvement in walking speed, but not in stride length. Patients of older age showed a somewhat higher improvement than younger patients after training with auditory cues. The significance of these results was enhanced by the results obtained for age-matched healthy controls, who showed no improvement in their walking abilities following training with either visual or auditory feedback (as the performance of these individuals can be assumed to leave little room for improvement, and the slight deterioration may be attributed to the burdening effect of the devices). Future studies may explore the long-term therapeutic effects of training with sensory feedback in a larger cohort of patients of various clinical subcategories, possibly in comparison to other treatment modalities. An extended-treatment course should be implemented and its effects examined in terms of both the specific functional performance and long-term skill learning. Research that can link sensory feedback-enhanced rehabilitation to long-term learning would have particular importance in CP, characterized by the damage of coordinated inter-neuronal networks leading to associated dysfunction. It might be hypothesized that gait improvement in CP patients employing sensory feedback is achieved through the enhancement of neuronal connections bypassing deficient pathways. However, the examination of this and other possible mechanisms for gait improvement and rehabilitation is left for further research. Sensory feedback-aided motor skill learning may also reveal interventional mechanisms for better resource allocation, functional reorganization, and the role of cerebral plasticity in neurorehabilitation.

The results were published in Baram and Lenger (2012).

Chapter 14

Virtual Reality Feedback for Gait Improvement in Patients with Idiopathic Senile Gait and Patients after Stroke

It has been demonstrated that sensory stimuli may improve the walking abilities of patients with movement disorders. In particular, earth-stationary visual cues were found to have a beneficial effect on gait in patients with Parkinson's disease (PD) (Martin, 1967; Azulay *et al.*, 1999). The closed-loop effects of real-world visual cues on the regulation and stabilization of gait were demonstrated in a previous study (Baram, 1999). This has led to the development of a wearable closed-loop virtual reality (VR) apparatus (Baram, 2004), which employs an inertial feedback system to incorporate the patient's body motions in the display, creating a visual sensation of moving with respect to earth-stationary tiles. The visual feedback device was found to be more effective for gait improvement in patients with PD than the open-loop version, displaying fixed-velocity motion (Baram *et al.*, 2002). The device was also found to improve gait in patients with multiple sclerosis (MS) (Baram and Miller, 2006). In the present study, this device was employed to study the effects of visual feedback cues on gait in patients with idiopathic senile gait (SG), some of whom had a previous history of strokes (PS) (Baram *et al.*, 2010).

The Israel Ministry of Health and the institutional Helsinki committees approved the study, and informed consent was obtained from each participant once the nature of the procedure was explained.

Twenty-one randomly selected old-age home residents suffering from SG, including six with PS, participated in the study. Exclusion criteria included considerable visual deficit not compensated for by correction, ocular movement dysfunction (e.g., diplopia, nystagmus), and gait disturbances due to pronounced muscle weakness, lower limb spasticity, sensory ataxia, or considerable general fatigue. Visual cues were generated by the closed-loop VR device shown in Fig. 14.1. Tests were performed at two old-age homes in Haifa, Israel, at approximately the same time in the morning. Examination of each patient comprised two stages, each consisting of the patient walking a straight 10-meter track four times. In Stage 1, the patient was verbally instructed to "walk normally" without the device. The time to complete the 10 meter track and the number of steps were recorded. In Stage 2, the device was put on the patient and turned on. The patient was instructed to walk on the virtual tiles while imagining walking on real tiles and attempting to reach the next tile in each step. The time to complete the 10 meter track and the number of steps were recorded. Walking speed (meters/second) and stride length (meters) were calculated from the time and the number of steps needed to complete the 10 meter track. The results were averaged for each stage.

Thirteen patients improved their walking speed, stride length, or both; nine improved their walking speed, stride length, or both by more than 10%. In patients with baseline performance above the median, the average improvement was considerably higher (6.3% ± 12.6% in walking speed and 6.4% ± 11.1% in stride length)

Figure 14.1. Visual-feedback virtual reality device used in tests.

than in patients with baseline performance below the median ($-0.7\% \pm 22.7\%$ in walking speed and $3.4\% \pm 11.3\%$ in stride length). Four of the six patients with PS improved their walking speed, stride length, or both by more than 10%. Although all four patients with left hemisphere vascular accident improved their gait, neither patient with right hemisphere vascular accident improved. In patients with PS with baseline performance above the median, improvement was considerably higher ($13.2\% \pm 6.0\%$ in walking speed and $16.6\% \pm 4.7\%$ in stride length) than in patients with baseline performance below the median ($-9.9\% \pm 27.0\%$ in walking speed and $-7.7\% \pm 12.3\%$ in stride length). The VR visual feedback cues did not improve gait in patients with vascular risk factors but without a history of PS. Education was a relatively good predictor of improvement (e.g., for the three patients with 8 years of schooling, average improvement was $-8.8\% \pm 23.8\%$ in walking speed and $-4.7\% \pm 15.3\%$ in stride length; for the four patients with 12 years of schooling, average improvement was $-1.8\% \pm 26.8\%$ in walking speed and $3.8\% \pm 9.7\%$ in stride length; and for the two patients with 20 years of schooling, average improvement was $6.7\% \pm 0.5\%$ in walking speed and $14.5\% \pm 12.6\%$ in stride length).

The results of the present study indicate that it is possible to use VR cues, superimposed on the real world in a closed-loop fashion, to help patients with senile gait and, in particular, patients with a history of strokes control their gait. The degree of improvement was found to be related to baseline walking speed and stride length and to level of education. Future studies should explore the long-term effects of training with such cues for therapeutic and rehabilitation purposes.

Chapter 15

Virtual Reality Cues for Improvement of Gait in Patients with Multiple Sclerosis

15.1 Introduction

Gait disturbances in patients with multiple sclerosis (MS) may result from pyramidal syndrome (muscle weakness, spasticity), sensory (proprio-kinesthetic perception) disturbances, and general fatigue, as well as incoordination due to cerebellar ataxia. About 85% of patients with MS develop gait problems (Armutlu *et al.*, 2001). Gait limitation results in mobility-related handicaps that can considerably limit the patients' activities of daily living, as well as their social interactions, and, therefore, have a major impact on the patients, their families, and society (Moster and Kesselring, 2002). To date, most studies on gait rehabilitation, whether by means of physiotherapy, occupational therapy, or pharmacologic treatment, have focused on improvement of muscle strength and reduction of spasticity, such as pharmacologic approaches using Baclofen (Nilsen *et al.*, 2000; Orsnes *et al.*, 2000; Thoumie and Mevellec, 2002; Lord *et al.*, 1998). However, the issue of cerebellar ataxia has gained little scientific attention, resulting in relatively poor treatment options.

Recent research has focused on the search for gait assessment methods in an attempt to lead to new approaches in intervention and rehabilitation. Typically, these studies consider gait in terms of time–distance measurements, i.e., speed and stride length, and observe their relation to muscle and joint movement analysis, muscular strength, and aerobic conditioning (Palliyath *et al.*, 1998; Benedetti, 1999; Morris *et al.*, 2002; Fauchard-Renard *et al.*, 2001; Mevellec *et al.*, 2003; Rodgers *et al.*, 1999). Patients with ataxia often use visual cues to maintain balance, which is the basis for the Romberg test. The application of virtual reality (VR) technology for the purpose of medical diagnosis and treatment has become widespread in the recent decade. Among its uses in neurologic rehabilitation is the development of motor assessment and training methods, using electromyographic, tactile (haptic), and visual cues (Ketalaer and De Wolf, 2000). A potential advantage of VR technology for the purpose of motor training is the capacity to provide the trainee with immediate performance feedback (Rizzo *et al.*, 2004).

It has been demonstrated that certain sensory stimuli improve the walking abilities of neurologic patients with movement disorders. In particular, moving visual cues, which generate a perception of motion, were found to have a beneficial effect on gait in patients with Parkinson's disease (PD) (Martin, 1967; Azulay *et al.*, 1999). A study of the closed-loop effects of real-world visual cues on the regulation and stabilization of gait (Baram, 1999) has led to the development of a virtual reality apparatus (Baram, 2004). The closed-loop design employs a sensory feedback system to incorporate the patient's body motions in the display. The use of this apparatus was found to be effective for gait improvement in patients with PD above and beyond the effects of the open-loop version (Baram *et al.*, 2002). In the present study, we employed an advanced version of this device to study the effects of visual feedback cues on gait in patients with MS. This study has been published, and further details, clinical and personal data of patients and control groups are presented in Table E-1 of Baram and Miller (2006) available in the *Neurology* website, www.neurology.org.

15.2 Methods

15.2.1 *Subjects*

Sixteen randomly selected outpatients with MS, seven men and nine women, diagnosed according to the Poser criteria (Poser *et al.*, 1983) with gait disturbances predominantly due to cerebellar ataxia, participated in the study. Exclusion criteria were considerable visual deficit not compensated by correction, ocular movement dysfunction (e.g., diplopia, nystagmus), and gait disturbances due to pronounced muscle weakness, spasticity, sensory ataxia, or considerable general fatigue. Disease-related disability was assessed using the Expanded Disability Status Scale (EDSS, Kurtzke). In addition, clinical assessment included Cerebellar Functional System Score (CFSS), which grades cerebellar function due to MS on a five-grade score: 0 = normal cerebellar function; 1 = abnormal signs without disability; 2 = mild ataxia; 3 = moderate ataxia; 4 = severe ataxia in all limbs or trunk; 5 = unable to perform coordinated movement due to ataxia.

15.2.2 *Ambulation assessment*

Assessment of patient ambulation, conducted by an impartial, independent investigator, was carried out using the ambulation measurements of walking speed (meters/second) and stride length (meters), calculated directly from measurements of time and number of steps needed to complete a fixed track. The control group consisted of 12 healthy individuals. The study was approved by the institutional Helsinki committee, and informed consent was obtained from each of the participants once the nature of the procedures had been fully explained. Clinical and personal data of patients and control groups are presented in Table E-1 of Baram and Miller (2006) available on the *Neurology* website, www.neurology.org.

Visual cues were generated by a virtual reality apparatus (Baram, 2004): a closed-loop head-mounted device shown in Figure E-1 of Baram and Miller (2006). The display, attached to the eyeglasses frame, provides the patient with a virtual tiled floor in a checkerboard arrangement, responding dynamically to the patient's own motion, much like a real floor, fixed in space.

15.2.3 *Procedure*

All tests were performed at the Multiple Sclerosis Center, Carmel Medical Center, Haifa, Israel, at about the same time in the morning. Examination of each patient comprised four stages, each consisting of the patient walking a straight track of 10 meters: baseline, online display off, online display on, and residual effects.

Stage 1: Baseline walking speed and stride length were measured first without the device. The patient was verbally instructed to start walking. The time to complete the 10-meter track and the number of steps were recorded four times and averaged.

Stage 2: The device was placed on the patient, but it was not turned on. Walking speed and stride length were again measured four times and averaged. The purpose of this stage was to produce a reference for the effect of the visual cue alone without the burdening effect of the device.

Stage 3: The display was turned on. The patient was instructed to walk on the virtual tiles along the 10-meter track. Walking speed and stride length were measured four times and averaged.

Stage 4: The device was taken off the patient, who was given a 10-minute break. After the break, the patient was instructed to walk the 10-meter track without the device. Walking speed and stride length were recorded four times and averaged. The purpose of this stage was to measure the residual short-term therapeutic effect of the visual cues.

15.3 Results

15.3.1 *Clinical features, baseline measurements, and online improvement*

The test results for the patients and the control subjects are summarized in Table E-2 of Baram and Miller (2006). Baseline walking speed and stride length of patients before using the device are given first. Walking speed and stride length of patients wearing the device, with the display turned off and on, are given in the "online" columns, along with percentage changes between "display off" and "display on". Walking speed and stride length after the break, without the device, are given in the "residual effect" columns, along with percentage changes with respect to the baseline walking speed and stride length.

The percentage improvement in the online walking speed of the patients and the controls with respect to the baseline walking speed is depicted for comparison in Fig. 15.1. Similarly, the residual improvement of the two groups is depicted in Fig. 15.2.

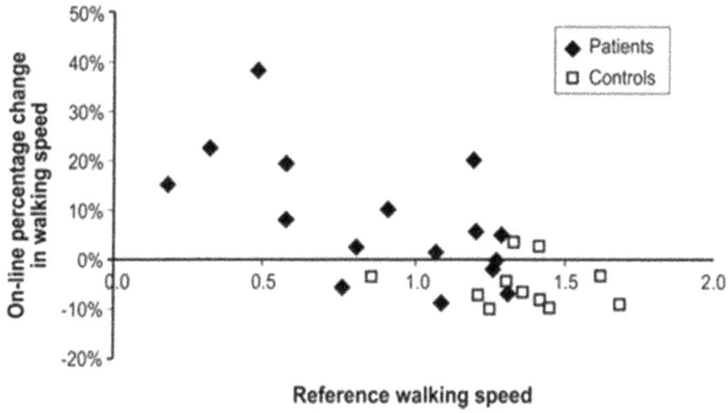

Figure 15.1. Online percentage improvement in the walking speed of patients with multiple sclerosis and of control subjects as a function of the baseline walking speed.

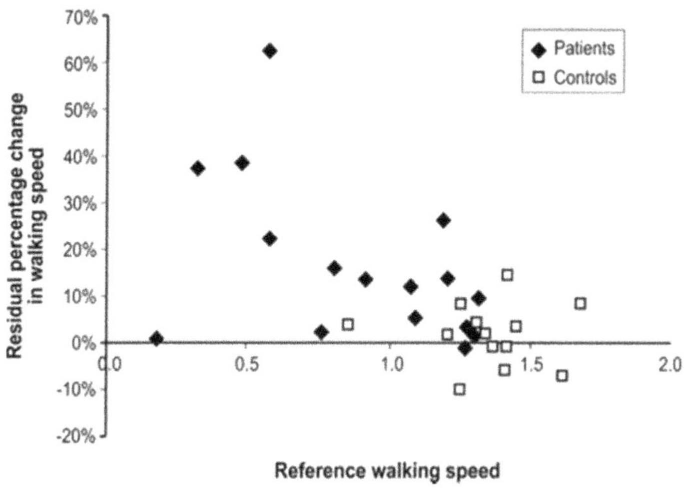

Figure 15.2. Residual (short-term therapeutic) percentage improvement in the walking speed of patients with multiple sclerosis and of control subjects as a function of the baseline walking speed.

The medians of the baseline walking speed and stride length were 0.994 m/s and 0.527 m for the patients and 1.379 m/s and 0.728 m for the controls.

For the patient group, the online walking speed improved, on average, by 13.46% in patients below the median and 1.47% in patients above the median. Using the baseline walking speed (BWS) value of 0.7 m/s as a separation threshold, it was found that the online walking speed improved, on average, by 20.32% in patients with BWS below the threshold and 1.63% in patients with BWS above the threshold. The online average improvement in stride length was 9.78% in patients below the median and 5.74% in patients above the median. Using the baseline stride length (BSL) value of 0.45 m as a separation threshold, it was found that the stride length improved, on average, by 14.01% in patients with BSL below the threshold and 4.92% in patients with BSL above the threshold.

In contrast to the results for the patients, the results for the controls did not show any particular trend in relation to the baseline walking speed or stride length. Furthermore, using the device did not improve performance; quite the contrary, due to the burdening effect of wearing the device. Indeed, it was found that the online walking speed changed, on average, by –5.3% in controls below the median and by –6.36% in controls above the median of the BWS. As in the case of the walking speed, there was an online decrease in stride length as a function of the BSL in the controls. The online average change in the stride length was also negative, –2.36% in controls below the median BSL and –1.08% in controls above the median. This makes the results for the patients even more noteworthy, since improved apparatus and prolonged training are expected to reduce the burdening effect and, hence, further improve performance.

15.3.2 *Residual improvement*

The residual improvement in the walking speed (see Fig. 15.2) was, on average, 24.49% in patients below the median BWS and 9.09% in patients above the median. Average residual improvement in walking speed was 32.61% in patients with BWS below 0.7 m/s and 9.59% in patients with higher BWS. In contrast, there was no notable residual improvement in the walking speed of the controls (2.49% and 2.36% changes in controls below and above the median BWS). The residual improvement in stride length of the patients was 16.32% in patients below the median BSL and 4.86% in patients above the median. Average residual improvement in stride length was 21.36% in patients with BSL below the 0.45 m threshold and 5.69% in patients with higher BSL. On the other hand, the controls did not show any residual improvement in the stride length (0.08% and –0.74% changes in controls below and above the median BSL).

During the tests, most of the patients reported a new sense of confidence and awareness. Following the tests, patients reported that an image of the tiles remains in their minds. One of the patients called a week after the test, saying that, when she walks, she is still walking on the tiles in her mind. Another patient reported in her subsequent visit that she walked better the whole week following the test.

15.4 Discussion

The advantage of VR implementation in neurologic rehabilitation is the ability to control and monitor the delivery of stimuli (Rizzo *et al.*, 2004). Applying this principle, VR-assisted methods of rehabilitation have been devised to assess and treat both cognitive deficits and motor disorders. A case in point is the use of VR to provide visual feedback cues to improve gait control in patients with PD. It has been suggested that the series of transverse lines may help trigger step transitions during walking in patients with PD (Martin, 1967), and the stabilizing effect of such information has been shown (Baram, 1999). Consequently, the delivery of a tiled floor pattern, responding dynamically to the patient's own motion, through a head-mounted VR device, has been suggested (Baram, 2004). When aided by the VR device, the gait quality of the patients with PD improved considerably (Baram *et al.*, 2002). The potential use of VR in the rehabilitation of patients with dysmetria, spasticity, and muscle weakness, using different stimuli, including haptics and EMG, has been proposed recently (Ketalaer and Wolf, 2000). The results of the present study clearly indicate that it is possible to use VR cues, superimposed on the real world in a closed-loop fashion, to help patients with MS control their gait. The degree of improvement, both online and residual, was found to be inversely related to the baseline walking speed and stride length. These improvements in performance seem particularly noteworthy when compared to the results obtained for the controls, which showed no meaningful improvement. Above all, our findings regarding residual improvement are particularly encouraging: average short-term residual therapeutic improvement in walking speed was 24.49% in patients with baseline walking speed below the median and 9.09% in patients above the median.

The results of the present study support the potential role of VR-based strategies as rehabilitation modalities in MS and substantiate their specific implementation in efforts to alleviate, improve, and restore mobility in patients with gait disturbances due to neurologic disorders in general.

Future studies may explore the implications of these results in a larger cohort of patients of various clinical subcategories (such as patients with above-median walking speed and below-median walking speed). An extended treatment course should be implemented and its effects examined in terms of both the specific functional performance and long-term skill learning. Research that can link VR-enhanced rehabilitation to long-term learning would have added importance in MS, characterized by the damage of coordinated interneuronal networks leading to associated dysfunction. Additionally, due to injury of CNS reserves in MS, cerebral reorganization and plasticity may be limited (Cifelli and Matthews, 2002). VR-aided neurorehabilitation that can demonstrate long-term motor skill learning may also elucidate important issues related to the modulation of neuronal networks for better resource allocation, the potential for functional cerebral reorganization, and the notion of interventional/adaptive cerebral plasticity as part of neurorehabilitation (Cifelli and Matthews, 2002).

Chapter 16

Auditory Feedback Control for Improvement of Gait in Patients with Multiple Sclerosis

16.1 Introduction

Positive effects of sensory cues on gait in patients with movement disorders have been reported before, particularly in the context of Parkinson's disease (PD) (Martin, 1967; Azulay *et al.*, 1999). Early attempts to produce such cues artificially have resulted in open-loop systems, producing sensory cues that are independent of the patient's own motion. Open-loop systems, subject to disturbances, are inherently unstable (Astrom, 1970). Indeed, such systems, displaying visual objects in constant motion (Prothero, 1993), have been found to cause confusion related to a sense of "falling out of sync" (Baram *et al.*, 2002). Open-loop rhythmic auditory cues have been found to produce different responses to different rhythms and raise issues of motivational factors (McIntosh *et al.*, 1997; Thaut *et al.*, 1996). In contrast, closed-loop visual feedback signals generated by the patient's own motion have been found to stimulate, stabilize and regulate gait (Baram, 1999). A study on patients with PD (Baram *et al.*, 2002) has shown that while open-loop stimuli have adverse effects, particularly dizziness, loss of balance, and even freezing, closed-loop visual feedback cues, responding to the patient's own movement, have a clear positive effect on gait. We have recently found a similar effect of closed-loop visual feedback on gait in patients with multiple sclerosis (MS), who suffer from cerebellar ataxia (Baram and Miller, 2007). Previously, auditory feedback has been found to have a positive influence on postural stability in quiet standing tasks (Dozza *et al.*, 2005). These findings suggest examining the effects of auditory feedback on gait in neurological patients. An apparatus that produces a ticking sound in response to the patient's steps, one tick per step, has been developed (Poser *et al.*, 1983). The patient hears the auditory cue produced by steps through an earphone and can modify the auditory cue by modifying gait. A steady, balanced gait will produce a pleasant auditory cue, synchronized with the patient's own steps, thus rewarding the patient for making the effort variable sequences of auditory feedback have been designed and tested first by Baram and published by Baram and Miller (2007).

16.2 Methods

16.2.1 *Subjects*

Fourteen randomly selected outpatients with MS, 10 women and four men, diagnosed according to the Poser criteria (Poser *et al.*, 1983) with gait disturbances predominantly due to cerebellar ataxia, participated in the study. Exclusion criteria include auditory dysfunction and gait disturbances due to pronounced muscle weakness, spasticity, sensory ataxia, or significant general fatigue. Disease-related disability was assessed using the Kurtzke Expanded Disability Status Scale (EDSS) (Kurtzke, 1993). In addition, clinical assessment included Cerebellar Functional System Score (CFSS) which grades cerebellar function due to MS on a five-grade score: 0 = normal cerebellar function; 1 = abnormal signs without disability; 2 = mild ataxia; 3 = moderate ataxia; 4 = severe ataxia in all limbs or trunk; 5 = unable to perform coordinated movement due to ataxia.

16.2.2 *Ambulation assessment*

Assessment of patient ambulation, conducted by an impartial, independent investigator, was carried out using the ambulation parameters of walking speed (meters/second) and stride length (meters). The control group consisted of 11 healthy individuals. The study was approved by the institutional Helsinki committee, and informed consent was obtained from each of the participants once the nature of

Table 16.1. Clinical characteristics of patients and control groups.

MS patient group						**Control group**		
Patient	**Sex**	**Age**	**DD**	**EDSS**	**CFSS**	**Subject**	**Sex**	**Age**
1	W	46	18	6	4	1	M	26
2	M	55	24	6	4	2	M	27
3	W	48	3	5	4	3	W	28
4	W	59	5	4	3	4	W	23
5	W	56	1.5	4	2.5	5	W	26
6	M	31	10	4.5	3	6	M	25
7	W	50	6	6	4	7	W	23
8	W	45	5	3.5	3	8	W	23
9	W	49	2	4.5	3	9	M	24
10	W	54	10	4.5	3.5	10	M	28
11	W	54	12	5.5	4	11	W	27
12	W	33	5	5.5	3.5			
13	M	49	21	3.5	2.5			
14	M	51	3	4	3			

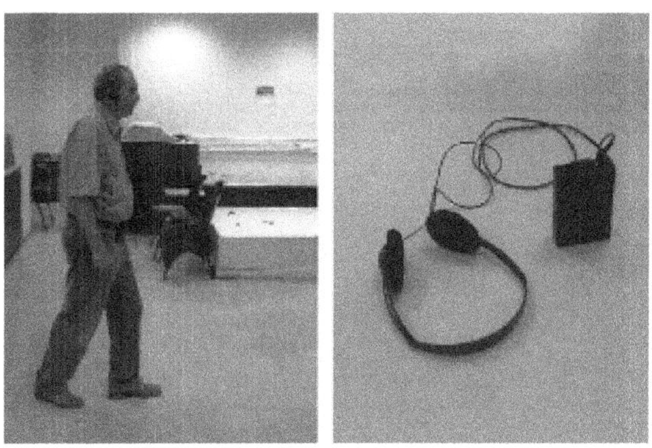

Figure 16.1. Auditory feedback apparatus used in tests.

the procedures had been fully explained. Clinical and personal data of patients and control groups are presented in Table 16.1.

16.2.3 *Auditory feedback apparatus and mode of assessment*

Auditory cues were generated by a device, shown in Fig. 16.1, which sounds, through a headset or an earpiece, a tick each time the user takes a step. A belt-mounted box, the size and weight of a small cell phone, contains a motion sensor, a microcontroller and software, which implements an adaptive filter for transforming the user's movement into sound. The user, in turn, adjusts gait so as to produce a balanced rhythmic auditory cue. The expected result is an improved gait pattern.

16.2.4 *Procedure*

All tests were performed at the Multiple Sclerosis Center, Carmel Medical Center, Haifa, Israel, at about the same time of day. Examination of each patient comprised four stages, each consisting of the patient walking a straight track of 10 meters: baseline, device off, device on, and residual effects, as explained hereinafter:

Stage 1: The ambulation parameters, baseline walking speed and stride length were measured first without the device. The patient was verbally instructed to "walk normally". The time to complete the 10-meter track and the number of steps were recorded four times and averaged.

Stage 2: The device was turned on. The patient was instructed to walk so as to make the auditory cue as rhythmic as possible along the 10-meter track. Walking speed and stride length were measured four times and averaged.

Stage 3: The device was taken off the patient, who was given a 10-minute break. After the break, the patient was instructed to "walk normally" the 10-meter track without the device. Walking speed and stride length were recorded four times and averaged. The purpose of this stage was to measure the residual short-term therapeutic effect of auditory feedback.

16.3 Results

16.3.1 *Clinical features, baseline parameters and online improvement*

The test results for the patients and the control subjects are summarized in Table 16.2. Baseline walking speed and stride length of

Table 16.2. Test results for MS patients and control subjects.

			Online				Residual			
	Baseline ambulation		Device on		Percentage change		Device off		Percentage change	
Patient no.	Walking speed (m/s)	Stride length (m)	Walking speed (m/s)	Stride length (m)	Walking speed (%)	Stride length (%)	Walking speed (m/s)	Stride length (m)	Walking speed (%)	Stride length (%)
MS patients										
1	0.252	0.256	0.400	0.333	58.73	30.01	0.361	0.312	43.25	21.86
2	0.647	0.444	0.493	0.435	−19.97	−15.20	0.684	0.526	5.72	18.47
3	1.043	0.513	0.984	0.541	−5.66	5.46	1.211	0.588	16.11	6.29
4	1.381	0.690	1.581	0.714	14.48	3.48	1.616	0.714	17.02	3.28
5	0.880	0.556	1.103	0.597	25.35	7.37	1.044	0.588	18.64	5.76
6	0.941	0.571	0.982	0.625	4.36	9.46	0.933	0.555	−0.85	−2.80
7	0.712	0.476	0.786	0.488	10.39	2.52	0.798	0.476	12.10	0.00
8	1.185	0.588	1.306	0.625	10.21	6.29	1.307	0.625	10.30	6.29
9	0.871	0.487	0.877	0.500	0.68	2.67	0.912	0.500	4.71	2.67
10	0.966	0.500	1.067	0.555	10.46	11.00	1.104	0.555	14.29	11.00
11	0.987	0.526	1.053	0.526	6.69	0,00	1.267	0.571	28.37	8.56
12	0.616	0.333	0.886	0.444	43.83	33,33	0.859	0.426	39.45	27.93
13	1.243	0.769	1.362	0.833	9.57	8.32	1.815	0.909	46.02	18.21
14	1.253	0.690	1.386	0.769	10.61	11.45	1.346	0.769	7.42	11.45

(Continued)

Table 16.2. *(Continued)*

Patient no.	Baseline ambulation		Online				Residual			
			Device on		Percentage change		Device off		Percentage change	
	Walking speed (m/s)	Stride length (m)	Walking speed (m/s)	Stride length (m)	Walking speed (%)	Stride length (%)	Walking speed (m/s)	Stride length (m)	Walking speed (%)	Stride length (%)
Control subjects										
1	1.361	0.769	1.298	0.769	−2.27	0.00	1.398	0.800	2.66	4.00
2	1.340	0.714	1.278	0.714	−6.58	0,00	1.407	0.714	5.07	0.00
3	1.852	0.833	1.953	0.833	2.54	0.00	1.890	0.833	2.08	0.00
4	1.576	0.714	1.500	0.769	−6.98	7.69	1.637	0.714	3.85	0.00
5	1.457	0.800	1.302	0.769	−12.83	−3.85	1.517	0.800	4.17	0.00
6	1.534	0.833	1.489	0.769	−1.04	0.00	1.572	0.769	2.52	−7.69
7	1.638	0.769	1.569	0.769	1.33	0.00	1.625	0.769	−0.81	0.00
8	1.408	0.667	1.420	0.690	−1.35	3.45	1.517	0.714	7.74	7.14
9	1.202	0.714	1.222	0.714	3.05	0.00	1.271	0.714	5.72	0.00
10	1.393	0.714	1.173	0.667	−5.92	0.00	1.284	0.667	−7.83	−6.67
11	1.308	0.667	1.311	0.667	−3.21	−3.33	1.329	0.667	1.59	0.00

patients before using the device are presented. Walking speed and stride length of patients using the device are specified in the "online" column along with percentage changes from "baseline" to "device on". Walking speed and stride length after the break, without the device, are given in the "residual effect" columns, along with percentage changes with respect to the baseline walking speed and stride length.

16.3.2 *Residual improvement*

The residual improvement in the walking speed (Fig. 16.3) was, on average, 18.75% (standard deviation: 18.53%). Average residual improvement in stride length was 9.93% (standard deviation: 9.46%). In contrast, there was no significant residual improvement in the walking speed and the stride length of the controls (2.43% and −0.29%, with standard deviations of 4.09% and 4.11%, respectively). Again, it can be seen that the level of variation is close to that of the average improvement.

The percentage improvement in the online walking speed of the patients and the controls with respect to the baseline walking speed is depicted for comparison in Fig. 16.2. Similarly, the residual improvement of the two groups is depicted in Fig. 16.3.

For the patient group, the online walking speed improvement was, on average, 12.84% (standard deviation: 18.74%). Online average

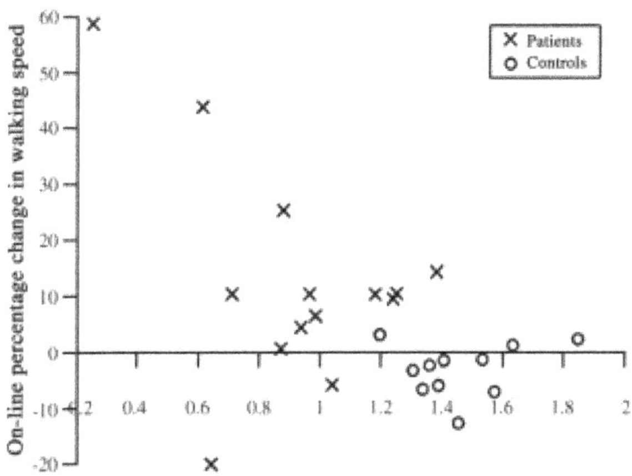

Figure 16.2. Online percentage improvement in the walking speed of MS patients and control subjects as a function of the baseline walking speed.

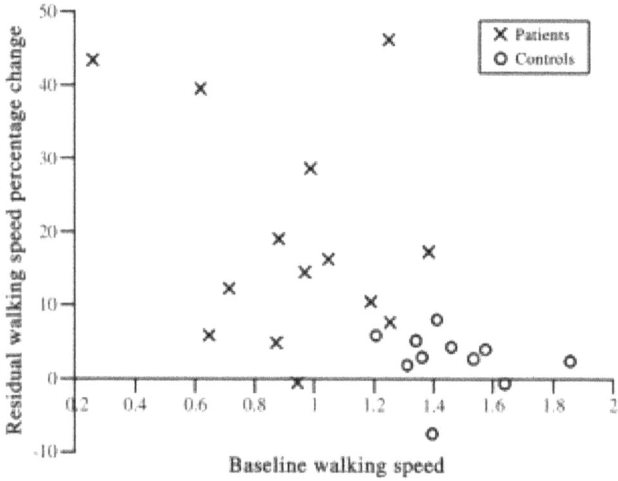

Figure 16.3. Residual (short-term therapeutic) percentage improvement in the walking speed of MS patients and control subjects as a function of the baseline walking speed.

improvement in stride length was 8.30% (standard deviation: 11.87%). The results for the controls did not show any particular trend in relation to the baseline walking speed or stride length. Furthermore, using the device did not improve performance due to the burdening effect of wearing the device. Indeed, it was found that the online walking speed changed, on average, by −3.02% (standard deviation: 4.76%). As in the case of the walking speed, there was no particular trend in the online percentage improvement in the stride length as a function of the baseline in the controls. The online average change in the stride length was also negligible (0.36%, with a standard deviation of 3.09%). It can be seen that the level of variation (or standard deviation) in the performance results among patients and controls is similar to the level of average improvement in performance.

16.4 Discussion

The mechanism by which auditory signals assist in coordinating rhythmic or sequential movements remains to be determined. There appears to be some evidence that rhythmic sound patterns can increase the excitability of spinal motor neurons via the reticulospinal pathway, thereby reducing the amount of time required for the muscles to respond to a given motor command (DelOlmo and Cuderio, 2003). There is also increasing evidence that the basal ganglia have an important role in the proper sequencing of complex movements (Palsev *et al.*, 1967). Studies on monkeys have shown that movement-related phasic discharge of pallidal neurons may serve as an internal cue to the supplementary motor area signaling the end of one movement and allowing the onset of the next (Brotchie *et al.*, 1991). These mechanisms do not necessarily hold the key to movement disorders in MS, where the problem is predominantly in the white matter and not in the gray matter. Furthermore, while the basal ganglia and the pallidum may play a role in certain movement disorders such as Parkinson's disease (Côté and Crutcher, 1991), the connection of the basal ganglia to ataxia, which is predominantly related to brainstem–cerebellar connections, is not clear. Yet, studies of neuronal substrates and mechanisms for auditory motor control may lead to strategies for gait management and rehabilitation in MS.

Open-loop verbal instructional cues (Behrman *et al.*, 1998) and rhythmic auditory cues (Thaut *et al.*, 1996; McIntosh *et al.*, 1997) have been used for gait management in patients with PD. However, open-loop rhythmic cues raise the question of which rhythm would suit a given patient at a given time (for instance, a rhythm 10% faster than the baseline (Baram *et al.*, 2002) may suit most, but not all patients). In addition, repeated use of rhythmic sound of relatively low complexity is assumed to induce a great amount of redundancy into the perceptual process and thus strongly reduce effective arousal effects related to motivation (Thaut *et al.*, 1996; McIntosh *et al.*, 1997). A study on the role of auditory feedback in maintaining postural balance in stance (Maurer *et al.*, 2003) suggests that auditory feedback increases postural stability in quiet standing tasks and results in a more prominent role for feedback (closed-loop) control over feedforward (open-loop) control. Furthermore, it indicates that the solution proposed by the brain with auditory feedback seems to involve more feedback control for a more stable sway. A comparison with visual feedback control of upright quiet stance (Rougier, 1999) suggests non-redundant roles in multi-sensory integration for the control of posture. Whereas vision provides information about the external environment and allows prediction of forthcoming events, auditory information processing time is markedly faster than visual reaction times, making it more important for postural reaction to disturbing stimuli (Rougier, 1999), i.e., stability feedback control. It should also be noted that an augmented auditory feedback device, employing a pressure-sensitive foot-switch, has been proposed for the treatment of equine (toe-walking) gait in children (Conrad and Bleck, 1980). Auditory feedback influences gait by creating an external cue, which serves as a reference for the patient. However, the feedback nature of this external cue makes it non-imposing, as it is fully controlled by the patient. It creates, on the one hand, a constant awareness of gait quality and, on the other, an instantaneous sensory response to changes in gait quality. The patient, by own control, then makes an effort to improve gait quality, and is, in turn, informed of any improvement (or deterioration) in gait quality by changes in the auditory cue.

The present study indicates that auditory feedback can help patients with MS control their gait. Patients with baseline performance below the median show a considerably higher improvement than patients above the median. In other words, the more room there is for improvement, the more improvement there is. Compared to our previous study (Baram and Miller, 2006), the present results show some notable differences between the effects of visual and auditory feedback. On average, gait improvement due to auditory feedback was higher than that achieved by visual feedback, which may require more elaborate information processing by the brain (Rougier, 1999). While visual feedback produced a higher improvement in stride length than in walking speed, the improvement in walking speed due to auditory feedback was higher than the improvement in stride length. The higher improvement in stride length caused by visual feedback can be attributed to the biofeedback effect of visually reaching a given target, i.e., the edge of a tile, by extending the foot farther. On the other hand, the higher walking speed caused by auditory feedback reinforces earlier findings that auditory signals reduce reaction time in a voluntary motor task (Conard and Bleck, 1980). Relative to the baseline performance, the residual improvement is even greater than the online improvement. This is explained by the fact that there was no offline training involved. Removing the burdening effect of the device, combined with the learning gained while using the device, maximizes the immediate residual performance.

The relatively large short-term residual improvement in patients using auditory feedback is particularly encouraging, as it suggests a therapeutic potential. The improvements in performance seem particularly noteworthy when compared to the results obtained for the controls, which showed no meaningful improvement. This suggests that an improved device, reducing the burdening effect, and extended training, may further improve patient performance.

Audio-based assistive technology may have special importance in the reinforcement of functions in patients who suffer from visual impairment (and are therefore not candidates for the visual system) and patients suffering from proprioceptive input problems. Since the auditory and the vision feedback channels seem to produce somewhat complementary functional enhancement properties, and since, as suggested by earlier studies (Rougier, 1999), multi-sensory integration may be key to motor stabilization and control, future studies may examine the effects of simultaneous visual and audio feedback through various artificial means on the walking abilities of MS and other neurological patients. Future studies may also examine the combined long-term therapeutic effects of visual and audio feedback on gait in movement disorder patients. The analysis of larger groups of patients would facilitate a comparison between patient subcategories, responders and non-responders to sensory feedback in particular, and a determination of specific predictive factors.

Chapter 17

The Gait-Entrainment Edge of Glide-Symmetric Visual Feedback in Multiple Sclerosis

17.1 Introduction

As noted in previous and subsequent chapters of this book, visual feedback has been found to affect the walking abilities of patients with neurological disorders. While early studies employed equally spaced transverse lines drawn on the ground (Martin, 1967), later reports noted similar effects by other visual patterns, such as a checkerboard-like tile arrangement (Sacks, 1973). In a recent study (Baram and Miller, 2006), we found that the improvement in the walking ability of multiple sclerosis patients training with a virtual reality device, which displayed earth-stationary checkerboard-like tiles, lasted beyond the use of the device. While this finding suggests that training with such cues may be an effective treatment modality for neurological gait disorders, the significance of the particular visual pattern used has remained an open question. Specifically, a question of interest is whether there is a fundamental difference between transverse lines and a checkerboard-like tile arrangement that might affect the level of gait improvement. An examination of the human locomotion mechanism reveals that it transforms the reflection symmetry of the limbs into a somewhat more compound symmetry called "glide reflection" (Livio, 2005).

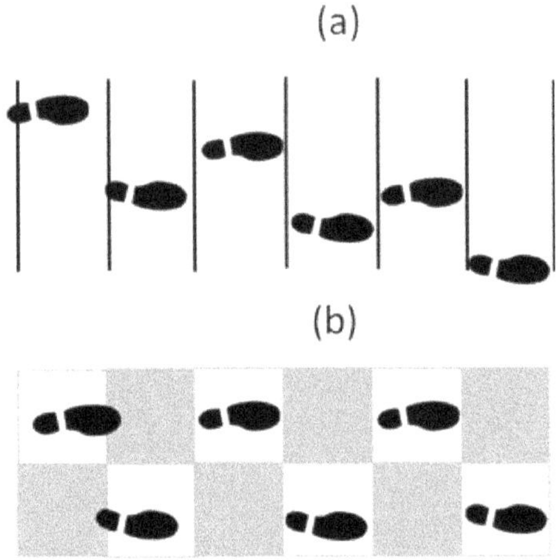

Figure 17.1. Walking with transverse lines (a) and with checkerboard tiles (b). Following a few initial steps, the steps synchronize with the lines or the tiles. The transition from one-dimensional (a) to two-dimensional (b) novelty and vigilance (see Chapter 10) results in lateral drift elimination.

As illustrated in Fig. 17.1(b), this geometry consists of both reflection and shifting. Unlike transverse lines (Fig. 17.1(a)) that have no distinct symmetry (although any line segment is reflection-symmetric about its middle point, there is no visible reflection surface), the geometry of the checkerboard tile arrangement (Fig. 17.1(b)) has a distinct glide-reflection symmetry. The checkerboard tile arrangement is matched then to he geometry of human locomotion.

In the context of walking, this geometry represents not only, as transverse lines, the reach of the leg in every step but also the transition from one leg to the other. We hypothesize that memorization of the visual feedback cue is a part of the gait training process, which would imply that training with the checkerboard tile arrangement will result in a higher level of improvement than training with transverse lines. This study was set to test this hypothesis.

17.2 Methods

17.2.1 *Subjects*

Twenty-one randomly selected MS outpatients, with gait disturbances predominantly due to cerebellar ataxia, participated in the study. Exclusion criteria include considerable visual deficit not compensated by correction, ocular movement dysfunction (e.g., diplopia, nystagmus), and gait disturbances due to pronounced muscle weakness, spasticity, sensory ataxia, or considerable general fatigue. Disease-related disability was assessed using the Kurtzke Expanded Disability Status Scale (EDSS, Kurtzke, 1993). The patients were randomly divided into two disjoint groups, one including 10 patients and the other including 11 patients. The first group was to be trained with visual cues consisting of transverse lines and the other with visual cues of a checkerboard-like tile arrangement.

17.2.2 *Ambulation assessment*

Assessment of patient ambulation, conducted by an impartial, independent investigator, was carried out using walking speed (meters/second) and stride length (meters), calculated directly from measurements of time and number of steps needed to complete a fixed track. The study was approved by the institutional Helsinki committee, and informed consent was obtained from each of the participants once the nature of the procedures had been fully explained.

17.2.3 *Virtual reality apparatus*

Visual cues were generated by a Virtual Reality apparatus (Baram *et al.*, 2002; Baram, 2004). The display, attached to the eyeglasses frame, provided patients of the first group with a display of transverse lines and patients of the second group with a virtual tiled floor in a checkerboard arrangement, both responding dynamically to the patient's own motion, much like a real floor, fixed in space. The side length of the tiles in the second arrangement was equal to the spacing between the transverse lines in the first.

17.2.4 *Procedure*

All tests were performed at the Multiple Sclerosis Center, Carmel Medical Center, Haifa, Israel, at about the same time in the morning. Examination of each patient comprised three stages:

Stage 1: Baseline walking speed and stride length without the device were measured. The patient was verbally instructed to walk along a straight track of 10 meters. The time to complete the track and the number of steps were recorded four times and averaged.

Stage 2: The device was placed on the patient and turned on. The patient was instructed to walk four times along the 10-meter track for the purpose of training. The patient was asked to imagine, while walking, stepping between the virtual lines or on the virtual checkerboard-like tiles, as applicable.

Stage 3: The device was taken off the patient, who was given a 10-minute break. After the break, the patient was instructed to walk the 10-meter track without the device. Walking speed and stride length were recorded four times and averaged. The purpose of this stage was to measure the residual short-term therapeutic effect of training with each of the two specified visual feedback cues.

17.3 Results

The test results for Groups 1 (training with transverse lines) and 2 (training with checkerboard tiles) are summarized in Tables 17.1 and 17.2, respectively, where clinical and personal data of patients are presented first. Baseline walking speed and stride length of patients before using the device are given in the "baseline" columns. Walking speed and stride length after training with the device are given in the

Table 17.1. Test results for patients of Group 1.

					Baseline		After training		Percentage change	
Patient	Sex	Age	DD (years)	EDSS	Walking speed (m/s)	Stride length (m)	Walking speed (m/s)	Stride length (m)	Walking speed (%)	Stride length (%)
1	W	59	7	3	0.755	0.426	0.808	0.488	7.02	14.55
2	W	33	4	4	0.820	0.455	0.870	0.488	6.10	7.25
3	M	44	19	3.5	1.196	0.714	1.422	0.769	18.90	7.70
4	W	39	14	5.5	1.140	0.606	1.252	0.625	9.82	3.14
5	M	16	1	3	1.186	0.690	1.259	0.690	6.16	0.00
6	W	51	6	2.5	1.191	0.645	1.277	0.667	3.02	3.40
7	W	21	1	3.5	1.211	0.588	1.257	0.645	3.80	9.69
8	W	45	10	5	0.497	0.417	0.525	0.426	5.63	8.67
9	M	39	6	3	0.707	0.465	0.772	0.500	19.19	7.53
10	W	43	12	5	1.168	0.606	1.264	0.667	8.22	10.07

Note: Training with transverse lines; DD = disease duration; EDSS = Expanded Disability Status Scale.

Table 17.2. Test results for patients of Group 2.

					Baseline		After training		Percentage change	
Patient	Sex	Age	DD (years)	EDSS	Walking speed (m/s)	Stride length (m)	Walking speed (m/s)	Stride length (m)	Walking speed (%)	Stride length (%)
1	W	24	5	6	1.093	0.555	1.153	0.588	5.49	5.95
2	W	20	5	4.5	1.316	0.645	1.440	0.667	9.42	3.41
3	M	54	23	6	0.577	0.425	0.708	0.513	22.70	20.71
4	M	30	2	5.5	1.072	0.625	1.204	0.667	12.31	6.72
5	W	34	3	4.5	0.808	0.476	0.962	0.513	16.32	7.77
6	M	50	1	3.5	1.194	0.667	1.508	0.769	26.30	15.29
7	M	67	4	6.5	0.178	0.162	0.180	0.172	1.12	6.17
8	W	50	25	2	1.272	0.625	1.278	0.606	−0.93	−3.04
9	W	48	6	5.5	0.480	0.357	0.666	0.417	38.75	16.81
10	W	46	18	3.5	0.321	0.290	0.442	0.370	37.69	42.31
11	W	47	3	5	0.575	0.385	0.936	0.465	62.78	20.78

Note: Training with checkerboard-like tiles; DD = disease duration; EDSS = Expanded Disability Status Scale.

"After training" columns, and the percentage changes with respect to the baseline walking speed and stride length are given in the "Percentage change" columns. The average improvement in Group 1 was 7.79% ± 4.24% in walking speed and 7.20% ± 3.92% in stride length. The average improvement in Group 2 was 21.09% ± 18.39% in walking speed and 12.99% ± 11.72% in stride length.

17.4 Discussion

It has been suggested that visual feedback of transverse lines may improve walking in patients with PD (Martin, 1967). Similar effects caused by tiled floor patterns have been reported (Sacks, 1973). Consequently, the stabilizing effect of such information has been shown (Baram, 1999; Baram *et al.*, 2002), and the delivery of a tiled floor pattern, responding dynamically to the patient's own motion, through a head-mounted VR device, has been suggested (Baram, 2004). The gait quality of PD patients (Baram *et al.*, 2002) and MS patients (Baram and Miller, 2006) aided by a portable virtual reality device (Baram, 2004) improved considerably. Yet, no particular significance has been associated with the geometry of the cue used for visual feedback. The results of the present study show that the choice of this geometry is of fundamental significance, having a substantial impact on the level of gait improvement. Specifically, a checkerboard-like tile pattern was found to have a considerably greater effect on gait improvement than a transverse line pattern. Furthermore, the level of improvement facilitated by the checkerboard-like tile pattern is inversely related to the level of impairment. We propose that the difference between the effects of the two patterns is associated with the types of symmetry that they convey. Unlike transverse lines, which have no distinct symmetry, the checkerboard tile arrangement has a distinct glide-reflection symmetry matched to the human locomotion mechanism. Besides the physiotherapeutic effect of gait training, the visual feedback generated by locomotion through the VR device appears to be memorized, controlling the orderly transition of the legs in a glide-symmetric fashion. This long-term effect makes training with glide-symmetric visual feedback a potentially effective treatment modality for patients with gait impairment.

Epilogue

In the years 1986–1988 and 1996–1998, I was invited to spend full sabbatical terms at the NASA Ames Research Center in Moffett Field, CA, which were followed by yearly summer visits in-between. Just before my travel back to Israel in 1998, I happened to watch American TV and was deeply moved by an interview with a Parkinson's disease patient. Returning to Israel, I kept thinking about that last experience. Waking up with that memory one morning, I felt I had a complete plan. It took me a few months to design a closed-loop portable audio-visual device, which translates human motion into vision and sound. Soon thereafter, I joined medical experts in Israel and other countries to perform clinical experiments on patients with Parkinson's disease, multiple sclerosis, idiopathic senile gait disorders, brain stroke, and cerebral palsy.

The following years were highly challenging and equally rewarding. A report submitted to the Technion president on May 7, 2000, by invited Turing award-winning professors, marked my research at the top of the computer science department achievements. I was invited and gave an oral presentation at the 53rd Annual Meeting of the American Academy of Neurology in Philadelphia, PA, visited by 12,000 medical experts, in May 2001. The experimental results were published in leading medical journals and presented in a variety of additional medical conferences. My device, approved by the US patent office in 2004, was found to be highly useful in helping patients suffering from different movement disorders. Having been invited to the main conference on multiple sclerosis in the summer of 2005, I lectured about my medical experiments and was awarded a plaque for the best lecture in that conference (the plaque is posted below). I was further invited as an honored guest at the next-year conference on multiple sclerosis. In 2006, I was appointed the incumbent of the Technion's Roy Mattas/Winnipeg Chair in Biomedical Engineering, and, shortly after, became head of the Technion Centre for Research in Intelligent Systems, until my retirement in November 2015. Yet, my medical and scientific interests, and a number of my additional publications, have followed me for years to come.

Finally, it might be noted that the title "Baram's Proximity Measure, (BPM)" was repeatedly mentioned by distinguished professor Michael Athans of MIT, in his plenary lecture and the attached manuscript he sent me before the 16th IFAC World Congress, Prague, Czech Republic, July 2005. His keynote lecture has repeatedly addressed the role of "Baram's Proximity Measure" in engineered mechanisms. It took me a few more years to realize that my original concept of proximity measures, established in my PhD thesis, is highly suitable for medical purposes. It should be emphasized that connected neural circuits were found, as shown in my previous World Scientific Books, "The Subcritical Brain (2021)" and "Motion Control (2023)", to play a major role not only in helping the neurologically impaired, but, by simple neural circuits correlation, in a large variety of brain activities.

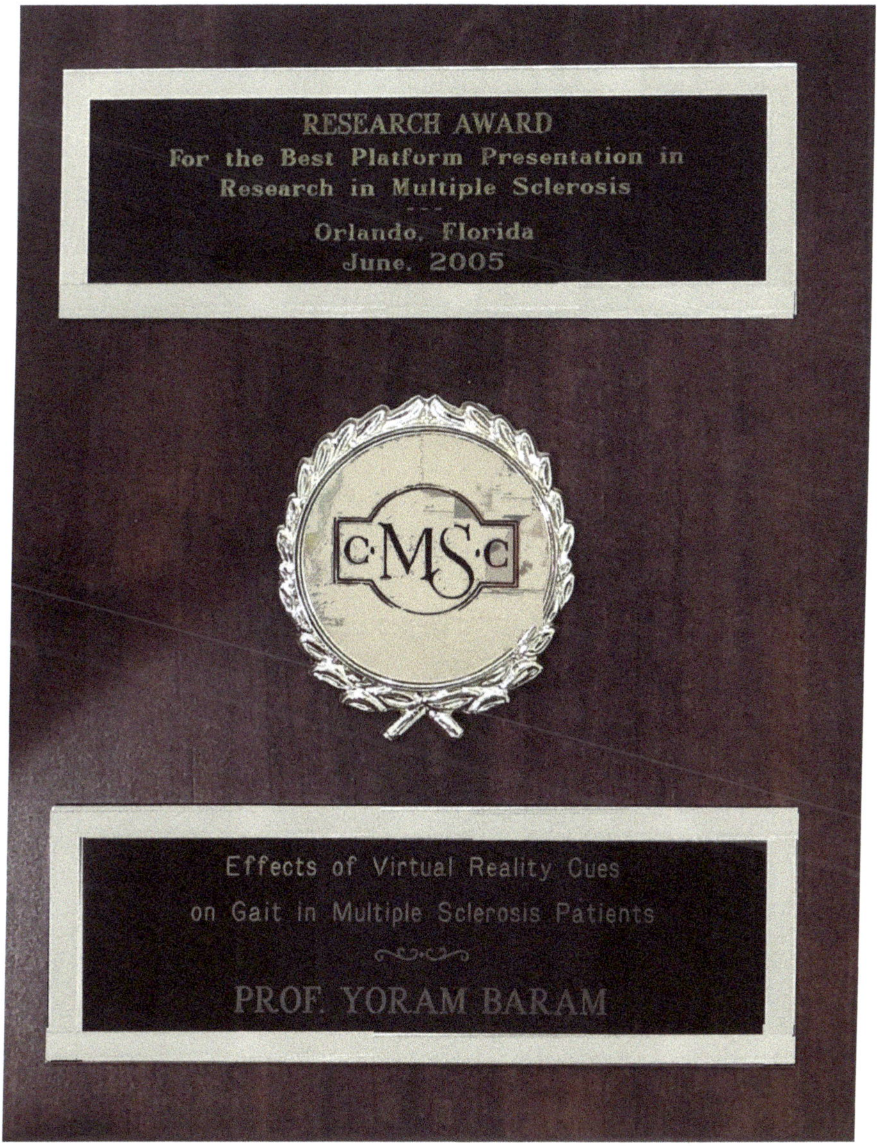
RESEARCH AWARD
For the Best Platform Presentation in
Research in Multiple Sclerosis
Orlando, Florida
June, 2005
C·MS·C
Effects of Virtual Reality Cues
on Gait in Multiple Sclerosis Patients
PROF. YORAM BARAM

References

Anderson, T. W. (1958). *An Introduction to Multivariate Statistical Analysis.* New York: Wiley.

Armutlu, K., Karabudak, R., and Nurlu, G., (2001). Physiotherapy approaches in the treatment of ataxic multiple sclerosis: A pilot study. *Neurorehabil. Neural Repair*, 15: 203–211.

Astrom, K. J. (1970). *Introduction to Stochastic Control Theory.* New York: Academic Press.

Athans, M., and Falb, P. L. (1966). *Optimal Control: An Introduction to the Theory and its Applications.* New York: McGraw-Hill.

Azulay, J. P., Mesure, S., Amblard, B., Blin, O., Sangla, I., and Pouget, J. (1999). Visual control of locomotion in Parkinson's disease. *Brain*, 122(Pt 1): 111–120.

Badarny, S., Aharon–Peretz, J., Susel, Z., Habib, G., and Baram, Y. (2014). Virtual reality feedback cues for improvement of gait in patients with Parkinsons disease. *Tremor and Other Hyperkinet Mov.*, 4: 225.

Bagley, S. Kelly, B., Tunniclife, N., *et al.* (1991). The effect of visual cues on the gait of independently mobile Parkinson's disease patients. *Physiotherapy*, 77: 415–420.

Baram, Y. (1996a). Classification by balanced representation. *Neurocomputing*, 13: 347–357.

Baram, Y. (1996b). A bird's eye view on the descent trajectory. *IEEE Trans. Aerospace Electron. Syst.*, 32(3): 1085–1087.

Baram, Y. (1999). Walking on tiles. *Neural Proc. Lett.*, 10(2): 81–87.

Baram, Y. (2004). Closed-loop augmented reality apparatus. *US Patent No. 6*, 734: 834–B1.

Baram, Y. (2012). Noninvertibility, Chaotic coding and chaotic multiplexity in synaptically modulated neural firing. *Neur. Comp.*, 24(3): 676–699.

Baram, Y. (2017a). Developmental metaplasticity in neural circuit codes of firing and structure. *Neural Netw.*, 85: 182–196.

Baram, Y. (2017b). Asynchronous segregation of cortical circuits and their function: A life-long role for synaptic death. *AIMS Neurosci.*, 4(2): 87–101.

Baram, Y. (2021a). Primal-size neural circuits in meta-periodic interaction. *Cogn. Neurodyn.*, 15(2): 359–367.

Baram, Y. (2021b). *The Subritical Brain.* World Scientific.

Baram, Y., and Sandell, NR Jr. (1978). An information theoretic approach to dynamical systems modeling and identification. *IEEE Trans. Automatic Control*, 23(1): 61–66.

Baram, Y., and Kailath, T. (1988). Estimability and regulability of linear systems. *IEEE Trans. Automatic Control*, AC-33(12): 1116–1121.

Baram, Y., and Barniv, Y. (1996). Obstacle detection by learning binary expansion patterns. *IEEE Trans. Aerospace and Electron. Syst.*, 32(1): 191–199.

Baram, Y., Aharon-Peretz, J., Simionovici, Y., and Ron L. (2002). Walking on virtual tiles. *Neural. Process. Lett.*, 16: 227–233.

Baram, Y., and Miller, A. (2006). Virtual reality cues for improvement of gait in patients with multiple sclerosis. *Neurology*, 66(2): 178–181.

Baram, Y., and Miller, A., (2007). Auditory feedback for improvement of gait in multiple sclerosis patients. *J. Neurol. Sci.*, 254: 90–94.

Baram, Y., Aharon-Peretz, J., and Lenger, R. (2010). Virtual reality feedback for gait improvement in patients with idiopathic senile gait disorders and in patient with history of strokes. *J. Amer. Geriatr. Soc.*, 58(1): 191–192.

Baram, Y., and Lenger, R. (2012). Gait improvement in patients with cerebral palsy by visual and auditory feedback. *Neuromodultion*, 15(1): 48–52.

Baram, Y., Aharon-Peretz, J., Badarny, S., Susel, Z., and Schlesinger, I. (2016). Closed-loop auditory feedback for the improvement of gait in patients with Parkinson's disease. *J. Neurol. Sci.*, 363(15), 104–106.

Bauer, H. (1972). *Probability Theory and Elements of Measure Theory.* Holt, Rineart and Winston, Inc.

Behrman, A. L., Teitelbaum, P., and Cauraugh, J. H. (1998). Verbal instructional sets to normalize the temporal and spatial gait variables in Parkinson's disease. *J. Neurol. Neurosurg. Psychiatry*, 65: 580–582.

Benedetti, M. G., Piperno, R.., Bonato, P., Tonini, A., and Giannini, S. (1999). Gait abnormalities in minimally impaired multiple sclerosis patients. *Mult. Scler.*, 5: 363–368.

Berger, T. (1971). *Rate Distortion Theory.* Prentice-Hall, Inc.

Berger, W. (1998). Characteristics of locomotor control in children with cerebral palsy. *Neurosci. Biobehav. Rev.*, 22: 579–582.

Berger, W., Quintern, J., and Dietz, V., (1982). Pathophysiology of gait in children with cerebral palsy. *Electroencephalogr. Clin. Neurophysiol.*, 53: 538–548.

Bhattacharyya, A. (1943). On a measure of divergence between two statistical populations defined by their probability distributions. *Bull. Calcutta Math. Soc.*, 35: 99–109.

Billingsley, P. (1961). Statistical methods in Markov chains. *Ann. Math. Stat.*, 32: 12–40.

Brotchie, P., Iansek, R., and Horne, M. K. (1991). Motor function of the monkey *Globus pallidus*. *Brain*, 114: 1685–1702.

Caines, P. E. (1975a). A note on the consistency of maximum likelihood estimates for finite families of stochastic processes. *Ann. Stat.*, 3(2): 539–546.

Caines, P. E. (1975b). Prediction error identification methods for stationary stochastic processes. *IEEE Trans. Auto. Contr.*, 21(4): 500–503.

Caines, P. E., and Rissanen, J. (1974). Maximum likelihood estimation of parameters in multivariate Gaussian stochastic processes. *IFFF Trans. Information Theory*, 20(1): 102–104.

Chacon, R. V., and Ornstein, W. (1960). A general ergodic theorem. *J. of Math.*, 4: 153–160.

Chen, Y.-P., Kang T.-Y., Doong J.-L., *et al.* (2007). Use of virtual reality to improve upper-extremity control in children with cerebral palsy: A single-subject design research report. *Phys. Ther.*, 87: 1441–1457.

Chung, K. L. (1974). *A Course in Probability Theory*. New York: Academic Press.

Cifelli, A., and Matthews P. M. (2002). Cerebral plasticity in multiple sclerosis: Insights from fMRI. *Mult. Scler.*, 8193–8199.

Colborne, G. R., Wright, F. V., and Naumann, S., (1994). Feedback of striceps surae EMG in gait in children with cerebral palsy. *Arch. Phys. Med. Rehabil.*, 75: 40–45.

Conrad, L., and Bleck, E. E. (1980). Augmented auditory feedback in the treatment of equinus gait in children. *Dev. Med. Child Neurol.*, 22: 713.

Côté, L., and Crutcher, M. D. (1991). The basal ganglia. In: Kandel, E. R., Schwartz, J. H., Jessel, T. M., editors. *Principles of Neural Science*, vol. 42. New Jersey: Prentice Hall. pp. 647–659.

Davies, M. N. O., and Green, P. R. (1990). Optic flow variables trigger landing in hawk but not in pigeons. *Naturwissenschaften*, 77(1990): 142–144, New York: Springer-Verlag.

DelOlmo, F., and Cuderio, J. A. (2003). Simple procedure using auditory stimuli to improve movement in Parkinson's disease: A pilot study. *Neurol. Clin. Neurophys.*, 2: 1–15.

Delval, A., Moreau, C., Bleuse, S., Tard, C., Ryckewaert, G., Devos, D., and Defebvre, L. (2014). Auditory cueing of gait initiation in Parkinson's disease patients with freezing of gait *Clin. Neurophysiol.*, 125(8): 1675–1681.

Doob, J. L. (1934). Probability and statistics. *Trans. Amer. Math. Soc.*, 36: 759–775.

Doob, J. L. (1953). *Stochastic Processes*. New York: John Wiley and Sons, Inc.

Dozza, M., Chiari, L., Chan, B., Rocchi, L., Horak, F. B., and Cappello, A. (2005). Influence of a portable audio-biofeedback on structural properties of postural sway. *J. Neuroeng. Rehab.*, 2: 13–28.

Dunne, J. W. Hankey, G. J., and Edi, R. H. (1987). Parkinsonism: upturned walking stick as an aid to locomotion. *Arch. Phys. Med. Rehabil.*, 68: 380–381.

Dursun, E., Dursun, N., and Alican, D. (2004). Effects of biofeedback treatment on gait in children with cerebral palsy. *Disabil. Rehabil.*, 26: 116–120.

Edgeworth, F. Y. (1908). On the probable error of frequency constants. *J. R. Stat. Soc.*, London, 71–72.

Espay, A. J. Baram, Y., Dwivedi, A. K., Shukla, R. M., Gartner, L., Gaines, A. P., Duker, F., and Revilla, J. (2010). At-home training with closed-loop augmented-reality cueing device for improvement of gait in patients with Parkinson's disease. *J. Rehabil. Res. Dev.*, 47(6): 573–582.

Espay, A. J., Li, J. Y., Johnston, L., Chen, R., and Lang, A., E. (2005). Mirror movements in parkinsonism: Evaluation of a new clinical sign. *J. Neurol. Neurosurg. Psychiatry.*, 76(10): 1355–1358.

Fahn, S., and Elton, R. L. (1987). Unified Parkinson's disease rating scale. In: Fahn, S. *Recent Developments in Parkinson's Disease.* New York: Raven Press, pp. 153–163.

Fano, R. M. (1961). *Transmission of Information.* New York: Wiley.

Fatou's Lemma (e.g., Bauer, 1972, p. 71).

Fauchard-Renard, C., Renard, J. F., Miret, N., Hannequin, D., Mihout, B., and Weber, J. (2001). Therapeutic efficacy during active phases of multiple sclerosis: Gait analysis and comparison with EDSS score. *Rev. Neurol. (Paris)*, 157: 649–654.

Ferrarin, M., Rabuffetti M., Tettamanti M., Pignatti R., Mauro A., and Albani G. (2008). Effect of optical flow versus attentional strategy on gait in Parkinson's Disease: A study with a portable optical stimulating device. *J. Neuroeng. Rehabil.*, 5: 3.

Fisher, R. A. (1922). On the mathematical foundation of theoretical statistics. *Phil. Trans. Royal Soc. London. Ser. A.* 222: 309–368.

Fisher, R. A. (1956). *Statistical Methods and Scientific Inference.* London: Oliver and Boyd.

Fletcher, W. A., and Aharpe, J. A. (1988). Smooth pursuit dysfunction in Alzheimer's disease. *Neurology*, 38(2): 272.

Gallager, R. G. (1968). *Information Theory and Reliable Communication.* New York: Wiley.

Gauss, C. F. (1809). *Theoria Motus Corporium Coeletum in Sectionibus Conicis Solem Ambientium.* Hamburg.

Giladi, N., McDermott M. P., Fahn, S., Przedborski, S., Jankovic, J., Stern, M., and Tanner C. (2001). Parkinson Study Group. Freezing of gait in PD: Prospective assessment in the DATATOP cohort. *Neurology*, 56(12): 1712–1721.

Giladi, N., Shabtai, H., Simon, E. S., Biran, S., Tal, J., and Korczyn, A. D. (2000). Construction of freezing of gait questionnaire for patients with Parkinsonism. *Parkinsonism Relat. Disord.*, 6(3): 165–170.

Griffin, H. J., Greenlaw, R., Limousin, P., Bhatia, K., Quinn, N. P., and Jahanshahi, M. (2011). The effect of real and virtual visual cues on walking in Parkinson's disease. *J. Neurol.*, 258(6): 991–1000.

Grossberg, S. (1993). A solution of the figure-ground problem for biological vision. *Neural Netw.*, 6(4): 463–483.

Grossberg, S., Mingolla, E., and Todorovoc, D. (1989). A neural network architecture for preattentive vision. *IEEE Trans. Biomed. Eng.*, 36(1): 65–84.

Grossberg, S., and Todorovic, D. (1988). Neural dynamics of 1-D and 2-D brightness perception: A unified model of classical and recent phenomena. *Percept. Psychophys.*, 43: 241–277.

Gupta, N. K., and Mehra, R. K. (1974). Computational aspects of maximum likelihood estimation and reduction in sensitivity function calculation. *IEEE Trans. Autom. Contr.*, 19: 774–783.

Gurland, J. (1954). On regularity conditions for maximum likelihood estimators. *Skand. Aktuar. T. Dskr.*, 37: 71–76.

Halmos, P. R. (1956). *Lectures on Ergodic Theory.* New York: Chelsea Publ. Co.

Hanakawa, T., Fukuyama, H., Katsumi, Y., Honda, M., and Shibasaki, H. (1999). Enhanced lateral pre-motor activity during paradoxical gait in Parkinson's disease. *Ann. Neurol.*, 45(3): 329–336.

Hawkes, R. M., and Moore, J. B. (1976). Performance bounds for adaptive estimation. *Proc. IEEE.*, 64(8): 1143–1150.

Hocherman, I., and Aharon-Peretz, J. (1994). Two-dimensional tracing and tracking in patients with Parkinson's disease. *Neurology*, 44(1): 111–116.

Hoehn, M. M., and Yahr, M. D. (1967). Parkinsonism: Onset, progression and mortality. *Neurology*, 17: 427–442.

Hutton, J. T., Nagel, J. A., and Loewenson, R. B. (1984). Eye tracking dysfunction in Alzheimer-type dementia. *Neurology*, 34(1): 99–102.

Jeffreys, H. (1946). An invariant form for the prior probability in estimation problems. *Proc. R. Soc. Lond. Ser. A*, 186: 453–461.

Jensen, J. L. W. V. (1906). Sur les fonctions convexes et les inégalités entre les valeurs moyennes. *Acta Math.*, 30(1): 175–193.

Kailath, T. (1967). The divergence and the Bhattacharyya distance measure in signal selection. *IEEE Trans. Comm. Tech.*, 15(1): 52–60.

Ketalaer, P., and De Wolf, L. (2000). New technologies in MS: Applications of virtual reality. *Int. MSJ*, 8: 11–19.

Klockgether, T., Borutta, M., Rapp, H., Spieker, S., and Dichgans, J. (1995). A defect of kinesthesia in Parkinson's disease. *Mov. Disord.*, 10(4): 460–465.

Koenderink, J. J., and van Doorn, A. J. (1975). Invariant properties of the motion parallax field due to the movement of rigid bodies relative to an observer. *Optica Acta*, 22(9): 773–791.

Kritikos, A., Leahy, C., Bradshaw, J. L., Iansek, R., Phillips, J. G., and Bradshaw, J. A. (1995). Contingent and non-contingent auditory cueing in Parkinson's disease. *Neuropsychologia*, 33(10): 1193–1203.

Kullback, S. (1959). *Information Theory and Statistics.* John Wiley and Sons, Inc. New York: Wiley.

Kurtzke, J. F. (1993). Rating neurological impairment in multiple sclerosis: An EDSS. *Neurology*, 33: 1444–1452.

Laplace, P. S. (1820). *Theorie Analytique de Probabilities.* 3rd ed. Paris. 309–354.

Lebesgue's dominated convergence theorem, (e.g., Chung, 1974, p. 421).

Lecam, L. (1953). On some asymptotic properties of maximim likelihood estimates and related Bayes' estimates. *Univ. Calif. Publ. Stat.*, 1: 277–330.

Ledebt, A., Becher, J., Kapper, J., *et al.* (2005). Balance training with visual feedback in children with hemiplegic cerebral palsy: Effect on stance and gait. *Motor Control*, 9: 459–468.

Lee, D. N., Davies, M. N. O., Green, P. R., and Van Der Weel, F. R. (Ruud). (1993). Visual control of velocity of approach by pigeons when landing. *J. Exp. Biol.*, 180(1): 85–104.

Lim, I., Van Wegen, E., De Goede, C., Deutekom, M., Nieuwboer, A., Willems, A., Jones, D., Rochester, L., and Kwakkel, G. (2005). Effects of external rhythmical cueing on gait in patients with Parkinson's disease: A systematic review. *Clin. Rehabil.*, 19(7): 695–713.

Liporace, L. A. (1971). Variance of Bayes estimates. *IEEE Trans. Inf. Theory*, 17(6): 665–669.

Ljung, L. (1974a). Convergence of recursive stochastic algorithms. Report 7403, Div of Auto Contr Lund Inst of Tech: February.

Ljung, L. (1974b). On consistency for prediction error identification methods. Report 7405, Div of Auto Contr Lund Inst of Tech: March.

Ljung, L. (1975). On consistency and identifiability. Report 7521, Div of Auto Contr Lund Inst of Tech: June.

Lord, S. E., Wade, D. T., and Halligan, P. W. (1998). A comparison of two physiotherapy treatment approaches to improve walking in multiple sclerosis: A pilot randomized controlled study. *Clin. Rehabil.*, 12: 477–486.

Mandl, F. (1988). *Statistical Physics.* 2nd Ed. Wiley and Sons.

Martin, J. P. (1967). Locomotion and the basal ganglia. In: Martin, J. P., editor. *The Basal Ganglia and Posture.* London (UK): Pitman Medical, pp. 20–35.

Maurer, C., Mergner, T., and Peterka, R. J. (2003). A multisensory posture control model of human upright stance. *Prog. Brain Res.*, 142: 189–201.

McIntosh, G. C., Brown, S. H., Rice, R. R., and Thaut, M. H. (1997). Rhythmic auditory-motor facilitation of gait patterns in patients with Parkinson's disease. *J. Neurol. Neurosurg. Psychiatry*, 62: 22–26.

Mevellec, E., Lamotte, D., Cantalloube, S., Amarenco, G., and Thoumie, P. (2003). Relationship between gait speed and strength parameters in multiple sclerosis [in French]. *Ann. Readapt. Med. Phys.*, 46: 85–90.

Mira, C. (2007). Noninvertible maps. *Scholarpedia*, 2(9): 2328.

Moore, J. B., and Hawkes, R. M. (1974). Performance of Bayesian parameter estimators for linear signal models. Report EE7410. Australia: University of Newcastle.

Moore, O., Peretz, C., and Giladi, N. (2007). Freezing of gait affects quality of life of peoples with Parkinson's disease beyond its relationships with mobility and gait. *Mov. Disord.*, 22(15): 2192–2195.

Morris, M. E., Iansek, R., Matyas, T. A., and Summers, J. J. (1994). The pathogenesis of gait hypokinesia in Parkinson's disease. *Brain*, 117(5): 1160–1181.

Morris, M. E., Iansek, R., Matyas, T. A., and Summers, J. J. (1996). Stride length regulation in Parkinson's disease: normalizations strategies and underlying mechanisms. *Brain*, 119: 551–568.

Morris, M. E., Cantwell, C., Vowels, L., and Dodd, K. (2002). Changes in gait and fatigue from morning to afternoon in people with multiple sclerosis. *J. Neurol. Neurosurg. Psychiatry*, 72: 361–365.

Nelson, R. C., and Aloimonos, J. (1989). Obstacle avoidance using flow field divergence. *IEEE Trans. Pattern Anal. Mach. Intell.*, 11(10): 1102–1106.

Nielsen, J. F., Anderson, J. B., and Sinkjaer, T. (2000). Baclofen increases the soleus stretch reflex threshold in the early swing phase during walking in spastic multiple sclerosis patients. *Mult. Scler.*, 6: 105–114.

Nieuwboer, A., Kwakkel, G., Rochester, L., Jones, D., Van Wegen, E., Willems, A. M., Chavret, F., Hetherington, V., Baker, K., and Lim, I. (2007). Cueing training in the home improves gait-related mobility in Parkinson's disease: The RESCUE trial. *J. Neurol. Neurosurg. Psychiatry*, 78(2): 134–140.

Orsnes, G. B., Sorensen, P. S., Larsen, T. K., and Ravnborg, M. (2000). Effect of baclofen on gait in spastic MS patients. *Acta Neurol. Scand.*, 101: 244–248.

Palliyath, S., Hallett, M., Thomas, S. L., and Lebiedowska, M. K. (1998). Gait in patients with cerebellar ataxia. *Mov. Disord.*, 13: 958–964.

Prazdny, K. (1981). Determining the instantaneous direction of motion from optical flow generated by a curvilinearly moving observer. *Comput. Vision Graph. Image Process.*, 17(3): 238–248.

Prothero, J. (1993). The treatment of akinesia using virtual images [master's thesis]. Seattle, WA: University of Washington, College of Engineering.

Poser, C. M., Paty, D. W., Scheinberg, L., McDonald, W. I., Davis, F. A., Ebers, G. C., *et al.* (1983). New diagnostic criteria for multiple sclerosis: guidelines for research protocols. *Ann. Neurol.*, 13: 227–231.

Ringach, D. L., and Baram, Y. (1992). Obstacle detection by diffusion. Technical report No. CIS–9220. Computer Science Dept., Technion, Israel Institute of Technology, November.

Ringach, D. L., and Baram, Y. (1994). A diffusion mechanism for obstacle detection from size-change information. *IEEE Trans. Pattern Anal. Mach. Intell.*, 16, 1: 76–79.

Riess, T., and Weghorst, S. (1995). Augmented reality in the treatment of Parkinson's disease. In: Morgan, K., Satava, R. M., Sieberg, H. B., Mattheus, R., Christensen, J. P., editors. *Proc. Medicine Meets Virtual Reality III.*, Amsterdam, The Netherlands: IOS Press. pp. 298–302.

Rissanen, J., and Caines, P. E. (1974). Consistency of maximum likelihood estimators for ARMA processes. Control Systems Report 7424 Det EE Univ Toronto: December.

Rizzo, A. A., Schulthheis, M., Kerns, K. A., and Mateer, C. (2004). Analysis of assets for virtual reality applications in neuropsychology. *Neuropsychol. Rehabil.*, 14: 207–239.

Rodgers, M. M., Mulcare, J. A., King, D. L., Mathews, T., Gupta, S. C., and Glaser, R. M. (1999). Gait characteristics of individuals with multiple sclerosis before and after a 6-month aerobic training program. *J. Rehabil. Res. Develop.* 36: 183–188.

Rougier, P. (1999). Influence of visual feedback on successive control mechanisms in upright quiet stance in humans assessed by fractional Brownian motion modelling. *Neurosci. Lett.*, 266(3): 157–160.

Roussas, G. G. (1965). Extension to Markov processes of a result by A. Wald about the consistency of maximum likelihood estimates. *Z. Wahrscheinlichkeitstheorie*, 4: 69–73.

Salimpoor, V. N., Zald, D. H., Zatorre, R. J., Dagher, A., and McIntosh, A. R. (2015). Predictions and the brain: How musical sounds become rewarding. *Trends Cogn. Sci.*, 19(2): 86–91.

Sacks, O. W. (1985). *The Man who Mistook his Wife for a Hat and Other Clinical Tales.* New York: Summit Books.

Savage, L. J. (1954). *The Foundations of Statistics*. New York: John Wiley and Sons.

Schelter, B., Winterhalder, M., Eichler, M., Peifer, M., Hellwig, B., Guschlbauer, B., Lucking, H. C., Dahlhous, R., and Timmer, J. (2006). Testing for directed influences among neural signals using partial directed coherence. *J. Neurosci. Methods*, 152(1–2): 210–219.

Seeger, B. R., Caudrey, D. J., and Scholes, J. R. (1981). Biofeedback therapy to achieve symmetrical gait in hemiplegic cerebral palsied children. *Arch. Phys. Med. Rehabil.*, 62: 364–368.

Stout, W. F. (1974). *Almost Sure Convergence*. New York: Academic Press.

Struik, D. J. (1969). *A Source Book in Mathematics 1200–1800*. Cambridge, MA: Harvard University Press. pp. 329–332.

Thaut, M. H., McIntosh, G. C., Rice, R. R., Miller, R. A., Rathbun, J., and Brault, J. M. (1996). Rhythmic auditory stimulation in gait training for Parkinson's disease patients. *Mov. Disord.*, 11(2): 193–200.

Thoumie, P., and Mevellec, E. (2002). Relation between walking speed and muscle strength is affected by somatosensory loss in multiple sclerosis. *J. Neurol. Neurosurg. Psychiatry*, 73: 313–315.

Tse, E. (1976). Identification and estimation of dynamic linear models with unknown structure. In: *Proc. JACC Conf.*, Purdue University.

Tse, E., and Weinert, H. L. (1975). Structure determination and parameter identification for multivariable stochastic linear systems. *IEEE Trans. Autom. Contr.*, 20(5): 603–613.

van der Heide, J. C., and Hadders-Algra, M. (2005). Postural muscle dyscoordination in children with cerebral palsy. *Neural Plast.*, 12: 197–203.

Vasilakos, K., and Beuter, A. (1993). Effects of noise on a delayed visual feedback system. *J. Theor. Biol.*, 165(3): 389–407.

Velu, P. D., Mullen, T., Noh, E., Valdivia, M., Poizner, H., Baram, Y., and de Sa, V. R. (2013). Effect of visual feedback on the occipito-parietal-motor network in Parkinsons disease patients with freezing of gait. *Front. Neurol.*, 4: 209.

Visser, H. (1983). Gait and balance in senile dementia of Alzheimer's type. *Aging*, 12(4): 296–301.

Wald, A. (1949). Note on the consistency of the maximum likelihood estimate. *Ann. Math. Stat.*, 20: 595–601.

Weghorst, S., Prothero, J., and Furness, T. (1994). Virtual images in the treatment of Parkinson's disease akinesia. In: *Proc. Medicine Meets Virtual Reality II*, San Diego, CA. pp. 242–243.

Weiner, N. (1949). *Extrapolation, Interpolation and Smoothing of Stationary Time Series*. Cambridge, MA: MIT Press.

Woodward, P. K. (1953). *Probability and Information Theory with Application to Radar*. New York: McGraw-Hill.

Woollacott, M. H., and Burtner, P. (1996). Neural and musculoskeletal contribution to the development of stance balance control in typical children and in children with cerebral palsy. *Acta Paediatr. Suppl.*, 416: 58–62.

Webster, K. E., Wittwer, J. E., and Feller, J. A. (2005). Validity of the GAITRite walkway system for the measurement of averaged and individual step parameters of gait. *Gait Posture*, 22(4): 317–321.

Zaccara, G., Gangemi, P. F., Muscas, G. C., Paganini, M., Pallanti, S., Parigi, A., Messori, A., and Arnetoli, G. (1992). Smooth-pursuit eye movements: alterations in Alzheimer's disease. *Neurol. Sci.*, 112(1–2): 81–89.

Index

About the Author

Yoram Baram of the Computer Science Department at the Technion, Israel Institute of Technology, received his BSc degree in Aeronautical Engineering from the Technion in 1972, MSc degree in Aeronautics and Astronautics from MIT in 1974, and PhD degree in Electrical Engineering and Computer Science from MIT in 1976. In 1978, he returned to Israel and joined Tel-Aviv University as a faculty member. He was soon joined by Prof. Uri Shaked in the discovery of fundamentally significant concepts of minimal-order system state estimation. In 1983, he moved to the Technion, Israel Institute of Technology, in Haifa, where he continued teaching undergraduate students and supervising graduate students in teaching the advanced theory of control systems.

In 1985, Baram was invited by Profs. Heinz Erzberger and Dallas Denery of the NASA Ames Research Center in Moffett Field, California, where he spent two sabbatical terms, in 1986–1988 and 1996–1998, as a Senior Research Associate of the US National Research Council. During his second sabbatical term, Baram was invited by Prof. Thomas Kailath of Stanford University for joint research, in which they discovered the fundamental concepts of *estimability* and *regulability*, which were previously unknown in control theory (Baram and Kailath, 1988). Before returning to Israel in 1988, Baram had found considerable interest in the newly emerging field of neural networks and, consequently, taught the first courses in neural networks at the Technion.

Baram's research and teaching during the years 1988–1998, including his second sabbatical term at NASA Ames, were devoted almost entirely to the concept of neural networks, addressing issues of memory capacity, vision, classification, pattern recognition, convergence, prediction, structure attraction, and trajectory selection. The latter concept, addressed by Baram's 1996 paper titled, "A bird's eye view on the descent trajectory", came as Baram's response to a question raised to him by Prof. Denery: "How can a pilot distinguish a three-dimensional object on the ground from a two-dimensional marking (such as an oil stain) when descending for landing on a runway"? A simple, mathematically defined approach trajectory is shown in that paper to safely resolve that dilemma.

Returning to Israel from his second sabbatical term in 1998, Baram has devoted himself to the development of a portable virtual reality device for safe and effective motion by neurologically impaired patients. Baram's preliminary work toward that goal was the first of only three works noted for their significance by the Evaluation Committee, appointed by the Technion President to assess the standing of the Faculty of Computer Science (Profs. R. Karp, and A. Pnueli, winners of the Turing award, and Prof. R. E. Bryant: The Technion Faculty of Computer Science, Report of the Review Committee, submitted to the President of the Technion on May 7, 2000).

The effectiveness of Baram's device in assisting gait and eliminating freezing in Parkinson's disease patients was demonstrated in collaborative studies with Prof. Judith Aharon-Peretz of the Technion Medical School and published in 2001. The results were revealed in a platform presentation by Baram at the 53rd Annual Meeting of the American Academy of Neurology, Philadelphia, 2001, a meeting attended by 12,000 neurologists from around the world. They were then widely publicized by the world media. The significance of Baram's device was that, in contrast to previously proposed approaches to motion improvement by the neurologically impaired, which were based on open loop, largely ineffective and dangerous to most patients as a result of error accumulation, gait entrainment by the closed-loop feedback device developed by Baram has, by virtue of feedback, an error correction effect, which protects every step of the patient from stumble, diversion or fall.

In 2004, Baram was awarded a US patent (No. 6,734,834-B-1, "Closed-loop Augmented Reality Apparatus") for his wearable virtual reality apparatus for gait improvement in neurological patients. In 2005, Baram received the Research Award for Best Platform Presentation in the leading conference on multiple sclerosis, CMSC, following joint work with Prof. Ariel Miller of the Technion Medical School. This work was published in 2006 in the leading journal, *Neurology.*

In early 2006, Baram was appointed the incumbent of the Technion's Roy Mattas/Winnipeg Chair in Biomedical Engineering. In mid-2006, Baram was invited by Prof. Alberto Espay to spend a sabbatical at the University of Cincinnati Medical School, where they conducted preliminary studies on the effects of Baram's sensory feedback device on gait in Parkinson's patients.

In 2012, Baram was joined by Dr. Ruben Lenger of ILAN — Israel Foundation for Handicapped Children in testing the ability of such children to walk safely with either the visual or the auditory virtual reality device on. The results in both cases, and residually 15 minutes after using either device, showed marked improvement in the walking abilities of most participants.

In 2010, Baram was appointed Head of the Technion Center for Intelligent Systems. His work, including further studies of motion impairment, involved mathematical disciplines such as chaotic attractors, random graphs, prime numbers and quantum computation, has been published in leading medical and neuroscience journals, and was summoned in Baram's two previous books, namely, *The Subcritical Brain: A Synergy of Segregated Neural Circuits in Memory, Cognition and Sensorimotor Control* (World Scientific, 2021), and *Motion Control: Multi-Faceted Movement in Space, Time and Neurological Impairment* (World Scientific, 2023).

www.ingramcontent.com/pod-product-compliance
Ingram Content Group UK Ltd.
Pitfield, Milton Keynes, MK11 3LW, UK
UKHW061933100926
8556IPUK00007B/901

9 789819 819263